D1219854

DRAWN FROM LIFE

Science and Art in the Portrayal of the New World

The use of images as evidence in historical writing has been largely neglected by historians, though recent interest in the importance of visualization in scientific literature has led to a reappraisal of their value. In *Drawn from Life*, Victoria Dickenson uncovers a vast pictorial tradition of 'scientific illustration' that reveals how artists and writers from the late sixteenth to the early nineteenth century portrayed the natural history and landscape of North America to European readers.

Dickenson undertakes a close reading of the images created by European artists, most of whom had never seen North America, and unravels the threads that linked the images to the curiosities and specimens that reached the Old World. Drawing on a wide range of illustrations – woodblock prints, engravings, watercolours, and maps – she examines several important issues regarding the nature of imagery: the tension between naturalistic representation and stylistic conventionalism; the role of the medium used in creating the image (especially the rise of printmaking); the historically changing function of images; and the need to consider historical context in 'reading' such pictures.

While many contemporary artists claimed that their work was 'drawn from life,' their images were, in fact, also works of the imagination. *Drawn from Life* is an illustrated archaeology of the imagination that allows readers to see North America as Cartier, Champlain, and early naturalists perceived it.

VICTORIA DICKENSON is Director of the McCord Museum in Montreal.

DRAWN FROM LIFE

Science and Art in the Portrayal of the New World

VICTORIA DICKENSON

UNIVERSITY OF TORONTO PRESS
Toronto Buffalo London

Toronto Buffalo London
Printed in Canada

ISBN 0-8020-4225-2 (cloth)
ISBN 0-8020-8073-1 (paper)

Printed on acid-free paper

Canadian Cataloguing in Publication Data

Dickenson, Victoria
Drawn from life : science and art in the portrayal of the New World

Includes bibliographical references and index.
ISBN 0-8020-4225-2 (bound) ISBN 0-8020-8073-1 (pbk.)

1. Scientific illustration – Canada – History. 2. Art and science – History.
3. Canada in art. I. Title.

Q222.D52 1998 502′.2′2 C98-931062-0

This book has been published with the help of a grant from the Humanities and Social Sciences Federation of Canada, using funds provided by the Social Sciences and Humanities Research Council of Canada.

University of Toronto Press acknowledges the financial assistance to its publishing program of the Canada Council for the Arts and the Ontario Arts Council.

CONTENTS

LIST OF PLATES

PREFACE

Most of my adult life has been spent in the world of museums. Even as a child, I enjoyed the company of objects, and passed many hours at the Royal Ontario Museum in Toronto, wandering its long galleries and wondering about the amazing things I saw. Objects inspired reflection. Who used this mirror, slept in this bed? How heavy were those swords, how difficult to wield? Objects, as some of the authors whom I cite will attest, are redolent with meaning, enabling us to query a past which they have shaped and been shaped by. Henry David Thoreau said that, when he looked at a collection of Indian arrowheads, he knew he was on the trail of mind.

I too have followed a trail in looking at the objects – chiefly works on paper – that form the basis for this study. I am interested in the minds and hands that created these images, and the eyes and minds that viewed them. My approach to this study of images is not that of the art historian. Unfortunately, I have not learned those powerful tools of analysis. But I have had many years' experience in looking, in what I refer to as 'close reading of objects.' I have applied this curatorial skill to the images discussed here, in an attempt to discern their meaning and their relations one with another, and in so doing to sketch the outlines of past perceptions.

In looking at the images, I have been concerned with their techniques of reproduction in a general sense. I am most interested in the process of translation from manuscript drawing to print. I am less concerned with identifying a print as an engraving or an etching than with knowing that it is an intaglio print on metal. In some cases, the techniques used by the artist or engraver are not easily discernible. George Edwards, for example, refers to his prints as gravings or engravings, when they are in fact etchings. In other cases, artists mix both techniques in creating the final print.

I have not confined my close reading to images, but have also read the texts which often accompanied them. Although I read many of the texts in French as well as English, I have chosen to quote from them in English, for the reader's convenience. There are excellent English translations of most of the French explorers' and travellers' accounts, some published at the time of the original French editions, others published more recently by organizations such as the Champlain Society. I have, however, left archaic spellings and constructions in many of the earlier quotations, since they are easily understood by most contemporary readers.

I have chosen, as well, to refer to animals and plants by their common English names, but have also included archaic spellings as well as Latin and French names, where appropriate. Some of the French terms incorporate unusual accents and spellings not currently in use.

Not all the images discussed in this study are illustrated with a plate, owing to the constraints of budget, as well as availability. In many cases, I have had to order specially the photographs of the maps, prints, and book illustrations. Many of the images discussed are also readily available in books and museum catalogues. Dürer's 'Rhinoceros,' for example, has been reprinted many times, from the sixteenth century to our own. As well-printed as the plates are, I would urge readers to seek out, where possible, original editions to appreciate better what people held in their hands and saw. Even if one is obliged to wear white cotton gloves, the feel of an old book is a fine thing.

ACKNOWLEDGMENTS

This study of images has taken place within the reserves and reading rooms of libraries, archives, and museums. My first debt is to those librarians, archivists, and curators who patiently unearthed material, showed me their treasures, and then allowed me to have them photographed. I would like in particular to thank the staff at the National Library of Canada, especially Dr Joyce Banks (now retired) and Peter Rochon, and the staff at the National Archives of Canada, particularly Gilbert Gignac, Eva Major-Marothy, and Ed Dahl. I was delighted to be given access to the vaults at the Botany Library at Agriculture and AgriFood Canada by Eva Gavora (also retired, and, like Dr Banks, not replaced). Eleanor MacLean and I spent hours together in the stacks of the Blacker-Wood Library of Biology at McGill University in Montreal, wearing white cotton gloves and leafing through the pages of the rare books in the library's quite extraordinary natural history collection. Bernadette Callery of the Library of the Bronx Botanical Garden in New York kindly lent me their 'big book,' Postels's massive tome on the *Vegetatio algorum*. Rex Banks (also retired), Librarian of the Natural History Museum in London, showed me Ehret's wonderful watercolours and a host of other materials not included here. Others who work in these institutions and other museums and archives responded to my calls, provided me with catalogue information, dug out prints and drawings, and inspired me with their dedication and their passion for the objects in their charge.

This study began as a catalogue for an exhibition at the Agnes Etherington Arts Centre in Kingston, Ontario, then, at the suggestion of Dr Beverly Boutilier, and with the support of Christopher Terry, Director General of the

National Aviation Museum, and my boss at the time, became a thesis. With the encouragement of my advisers at Carleton University, particularly Dr Brian McKillop, and the editors at the University of Toronto Press, from Virgil Duff to Gerald Hallowell and Emily Andrew, and finally Darlene Zeleney, the thesis has been translated into a book. My thanks to the Press, and to my careful copy-editor, Beverley Beetham Endersby, and the book's designer, Val Cooke, for their felicitous conversion of typescript to printed page.

Finally, it is impossible to acknowledge adequately the support and encouragement of my husband, Jeffery Harrison, and our three children, Peter, Luke, and Adam Harrison. They have put up with me tired, cranky, and irritable, and have cheerfully assumed that it must be worth it. This book truly is a tribute to their good humour and appreciation of the world of ideas.

PLATE 1: George Edwards (1694–1773), ***The Bittern from Hudson's-Bay***, 1748. Etching, hand-coloured. From George Edwards, *A Natural History of Uncommon Birds ...*, part III, plate 136.

PLATE 2: Pierre Desceliers (1487–1553), Map of North America, from ***A Map of the World***, 1546. Colour lithograph (original watercolour). Nineteenth-century redrawing of the original, taken from *Les Monuments de la géographie* by Edme François Jomard (Paris, 1854). The original map is in the John Rylands University Library of Manchester, England. The convention that north should appear at the top of the map had not been established firmly, and thus the map appears to modern eyes to be 'upside down.'

PLATE 35: George Edwards (1694–1773). Frontispiece, 1743. Etching, hand-coloured. From George Edwards, *A Natural History of Uncommon Birds* ..., part I.

PLATE 39: Peter Mazell (fl. 1761–1797), after George Stubbs (1724–1806), Moose deer, 1792. Hand-coloured engraving. From Thomas Pennant, *Arctic Zoology* (London, 1792), VIII. George Stubbs was hired in 1770 by William Hunter to paint an oil portrait of the Duke of Richmond's yearling bull moose. The portrait was exhibited at the Society of Artists in 1783, and Pennant requested that Mazell be allowed to prepare an engraving for inclusion in his book. This engraving is hand-coloured in Pennant's personal copy of the 1792 edition.

Drawn from Life

INTRODUCTION: *THE BITTERN FROM HUDSON'S-BAY*

It is almost twenty years since I first encountered a print signed 'Geo. Edwards,' of a bittern 'from Hudson's-Bay' (plate 1; see colour plates). The print – a copperplate etching, hand-coloured and dated 1748 – was obviously a page from a book. The bird itself looked strange; the plumage was reasonably correct, but the stance awkward and unnatural. On an accompanying text page, the author had compared this bird, brought from '*Hudson's-Bay* by Mr. *Isham*,' with one taken near London, and found that it was somewhat smaller than the European bird. 'It is very much the Colour and Make of our Bittern, and hardly to be known from it but by Comparison.' The author identified the bird as a nondescript species, and suggested that 'the Curious who would compare this Description with that of our common Bittern may see it in *Willughby's* Ornithology, *P.* 283.'

Who were Edwards, Willughby, and Isham? Why was the description so detailed, the drawing so imperfect? How was the bird brought from North America? In what book had this print appeared? What was the book's purpose? Who read it? Since I first saw *The Bittern from Hudson's-Bay*, I have come across other drawings of birds done by George Edwards – ill-proportioned nighthawks perched in trees, peculiar ducks, strange flycatchers. I have also read his books and those of other eighteenth-century author-naturalists, such as Mark Catesby, Eleazar Albin, and Thomas Pennant. I have discovered a tradition of natural-history illustration and publication which began in the sixteenth century and continues today. I have also answered some of my original questions, and found others, and both my answers and my speculations form the basis of this book.

Edwards's plate of the bittern appeared in his *A Natural History of Uncommon Birds ...*, published in London between 1743 and 1751. George Edwards (1694–1773) was the bedell (or beadle) of the College of Physicians, a friend and colleague of Sir Hans Sloane's (1660–1753), and a naturalist and artist. He travelled widely in Europe but never visited North America. He published another book, *Gleanings of Natural History* (1758–64), which also included some North American species. Francis Willughby (1635–1672) was a seventeenth-century English naturalist, who, with John Ray (1627–1705), travelled extensively in Britain and northern Europe, and prepared an *Ornithology*, published posthumously in 1676. James Isham (*ca* 1716–1761) was an employee of the Hudson's Bay Company and supplied some North American specimens to collectors such as Sir Hans Sloane. He was acquainted with Edwards, who praised his 'commendable curiosity.'[1] The bittern itself had probably been sent to England in a Company ship, either as a skin, or preserved whole and salted. Edwards was both artist and printmaker, an unusual dual role. He had been taught the art of 'Etching on Copper with Aqua-Fortis' by a fellow artist, Mark Catesby, who had also taught himself, the cost of engravers being beyond the resources of both men. Edwards was also a self-taught artist, having decided in his twenties to pursue natural-history studies full-time. His biographer notes that, on his return from his European travels in 1719,

> Mr. Edwards closely pursued his favourite study of Natural History; applying himself to drawing and colouring such animals as fell under his notice. A strict attention to natural, more than picturesque beauty claimed his earliest care: Birds first engaged his particular attention; and having purchased some of the best pictures of these subjects, he was induced to make a few drawings of his own; which were admired by the curious, who encouraged our young naturalist to proceed, by paying a good price for his early labours.[2]

Edwards was admired, respected, even eulogized for his skills in depicting animals in their 'natural' beauty. His books appear to have sold well, some editions sporting bilingual French and English text, other editions appearing in English and in German. He was praised as an artist and a naturalist, and yet to the modern eye Edwards's illustrations are stilted, even inept. George Edwards was not, however, judged by his peers and colleagues a poor or incompetent artist. Linnaeus considered his works on birds one of 'the miracles of our century,' equal to Georg Ehret's botanical drawings.[3] Linnaeus also noted in a letter that Edwards's drawings were so accurate that 'nothing is wanting to the birds but their song.'[4] Sir Hans Sloane, whose collections formed the foundation of the British Museum and who was a great savant, hired Edwards to add to his drawing collection, procured for him an agreeable sinecure, and remained his patron.

There is here, between ourselves and eighteenth-century authors, readers, and naturalists, a dissonance, a division which cannot be blamed on problems of skill, reproductive techniques, poor printing, or marginal interests. What is the nature of the dissonance – what has changed? What can an examination of works like Edwards's *Bittern* or other illustrations of flora, fauna, and natural features tell us about science, art, and the view of life held in the past? My inquiry starts, then, with the image, not the word. I begin by looking closely at the *Bittern*, and a selection of the multitudes of images of birds and animals, insects and fish, flowers and trees, waterfalls and rock formations that we now refer to as 'scientific illustration.' These images were drawn primarily for the purpose of communicating information, and were included on maps and printed in herbals and florilegia, and in books of discovery and exploration, technical manuals, and natural histories. For the most part, we have forgotten these old images of the natural world, which for centuries flooded from the brushes, pens, and presses of Europe. They have disappeared either under the tide of landscape painting, which in the eighteenth and nineteenth centuries subsumed other depictions of the landscape, or under the impact of photography, with its ability to render more detail than the eye can see with casual observation. That these images fulfilled some significant need in the intellectual world cannot be doubted. Their very ubiquity is testimony to the important role they had in the shaping of knowledge, a role that has long been neglected by historians. At the same time, the conventions and visual language they employed, which appear foreign to contemporary eyes, suggest that the understanding of the informational value of the image has changed, or at least that the information we require from images differs from what was required by scientific thinkers and knowledgeable readers in the past.

I have decided to examine a certain set of these images, those which depict some aspect of the natural landscape, or biota, of North America, and most particularly its northern half, from the discovery period, in the fifteenth century, to the early nineteenth century. The choice of this limited set was necessitated both by the masses of material available for study and by the accessibility of original materials.[5] The decision to include only North American or, where possible, 'Canadian' images also permits not only an examination of the role of images in the shaping of knowledge, but also a chance to comment on the particular vision of the northern half of the New World held by Europeans from the age of discovery and settlement to the nineteenth century. These are also, for the most part, the images of my own landscape, of the flora and fauna with which I have a more-than-passing familiarity. The skills required to identify the types of plants or animals we meet in daily life are not highly prized by the majority of North Americans, though there exists a significant minority, born in part from the environmental movement and in part from the enduring passion for natural history that has existed since the eighteenth century, who take pleasure in the ability to name the other members of the local biota. The naturalist's passion – the naming of things – is more than simple listing. It is, I think, primarily an attempt at ordering, assigning to a place, not as a mechanism of control, but as a heuristic for understanding. Understanding is required only when we assign significance to phenomena. One of the areas upon which this work touches, then, is the shifting of interest from the natural world as metaphor, with a limited number of icons, to the natural world as the real 'other,' with a multitude of particular species.

For many early writers in the fifteenth and sixteenth centuries, the contents of the world – the plants, animals, minerals, and phenomena that they could enumerate – had changed little since Pliny the Elder (23–79) first recorded his observations in his *Natural History* in the first century. Their knowledge of the natural world, or at least the knowledge they chose to record, reflected the interests of the Roman world, detailing the plants and animals of the Mediterranean littoral and the exotica of the Near East. The new plants and animals that the voyages of exploration revealed to the Old World changed the content and the value of natural history. Challenged with describing the indescribable (or at least the unfamiliar), the naturalists of the late sixteenth and seventeenth centuries and their followers began to enumerate the inhabitants of the world in a manner that differed considerably from that of their intellectual forebears. The fascination with the new and the strange led to an understanding of the diversity of life, the story of which is still being explored today by contemporary biologists and palaeontologists.

To return to the importance of knowing the local biota, my own acquaintance with the birds, animals, plants, and scenes described by George Edwards and his colleagues, and by other writers and artists before and after them, enables me to exercise some kind of critical judgment in examining these images. By this, I mean I can apply my own knowledge of the subjects seen in the wild to the understanding of the representations made by the artists, who sometimes worked under considerable disadvantage in their efforts at rendering an image drawn from nature. This is not to imply that I assume all representations should exhibit a late-twentieth-century standard of photographic realism (indeed, much of contemporary natural-history illustration, particularly in field guides, does not), but that familiarity with the appearance of the actual animal or plant, or at least an individual specimen, is an advantage in understanding both the constraints under which the artists worked and the particular perspective they brought to the problems of depiction.

How does one begin to pick away at these images, to unravel the threads which connect them to each other and to the world in which they were created? The first, and perhaps most important task, is to treat them seriously. In 1976, Martin Rudwick pointed out in his article 'The Emergence of a Visual Language for Geological Science, 1760–1840'[6] that 'in modern historical analysis the strong visual component of the original source-materials is generally either missing altogether or else reduced to a virtually decorative role.' He attributes this lack primarily to the absence in the history of science of 'any strong intellectual tradition in which visual modes of communication are accepted as essential for the historical analysis and understanding of scientific knowledge.'[7] This neglect of the visual mode is obvious in an examination of reprinted materials. Where the original work might have included a decorative and allegorical frontispiece, or several illustrations, new editions are often limited to text, as if only the text were freighted with meaning, and the illustrations were in fact meaningless, or at best attractive marginalia. At the opposite extreme, there exist a number of modern works in which images from earlier printed books, even details from larger pictures, are extracted from their original context and used as decorative footnotes to text, abrogating the contextual-information value of the image.[8] This misuse of images, even by historians, is widespread and is complicated further by an often casual attitude to chronology. Any old picture of a bear can illustrate any old text about a bear, even when the image is seventeenth-century and the text sixteenth-century. This disregard for chronology would not be tolerated in the treatment of text, where strict temporal order is regarded as a tool of historical analysis. It is as if the world of vision and the world of discourse could be separated at will, as if they had no connection with each other, and as if thinking proceeded by word alone. Brian Baigrie, in the introduction to *Picturing Knowledge: Historical and Philosophical Problems Concerning the Use of Art in Science*, suggests that part of the reason for this cavalier treatment of images is 'the deep-seated conviction on the part of science studies scholars that human thinking takes place in words, [and] the supposition persists that pictures in science are psychological devices that serve as heuristic aids when reasoning breaks down.'[9] There is considerable evidence to suggest that thinking is a far more complex activity than the forming of mental sentences, and the role of picturing in thought and its relationship to verbal description is only beginning to be appreciated.[10] Thinking seriously about the role of images, then, means paying attention to the interpenetration of text and image, and to chronology. Examining the way in which images and texts are interwoven offers new and clearer perspectives on the use of imagery and its role in thinking about the natural world.

This split between the world of image and the world of text is exemplified by an examination of recent printed sources for representations of natural-history subjects. They appear primarily in the catalogues of museums, archives, and libraries, and in the works of art historians. In the world of museums, the painted or printed illustration or the illustrated rare book is classified in much the same way as an artefact – as a unique thing. Prints and drawings departments of museums, in which many of the images of 'documentary art' are held, tend to be staffed by art historians, whose interest in the image is often ahistorical, in the sense that they ascribe the greatest significance to the aesthetic value of the image or to its place in the oeuvre of an artist or a school. The same is often true of image collections held in libraries or archives, where prints, drawings, maps, and rare books tend to be separated from the book collections, and treated as supplements to the actual tools of learning – that is, the manuscripts and the printed books. Fortunately, the appeal of these images, and the interest which they hold for both specialized and general audiences, have resulted in the publication of lavishly illustrated volumes that make the collections more accessible. A selection of such works published since the late 1970s includes *Flowers in Books and Drawings, ca 940–1840*, published by the Pierpont Morgan Library (1980); *Views and Visions: American Landscape before 1830*, from the Corcoran Gallery in Washington (1986); *The European Vision of America*, from the Cleveland Museum (1975); *The Bison in Art*, curated by the Amon Carter Museum in Fort Worth (1977); *Encountering the New World, 1493 to 1800*, a catalogue for the Columbus Quincentenary by the John Carter Brown Library (1991); the National Gallery of Art's magisterial *Circa 1492: Art in the Age of Exploration* (1991); *Images of Science: A History of Scientific Illustration*, primarily from the collections of the British Library (1992); and *Passages*, from the collections of the National Library of Canada (1992).[11]

While the historical study of visual communication in science has often been neglected by historians of science, art historians have a long tradition of interest in original and printed illustrations of science and nature. Otto Pacht wrote a pioneering article in 1950 in the *Journal of the Warburg and Courtauld Institutes* titled 'Early Italian Nature Studies and the Early Calendar Landscape,'[12] while, in 1971, Francis Klingender published *Animals in Art and Thought to the End of the Middle Ages*. Svetlana Alpers related the work of seventeenth-century Dutch artists to the science of their day in her controversial 1983 book, *The Art of Describing: Dutch Art in the Seventeenth Century*. Barbara Maria Stafford looked at travel literature in *Voyage into Substance: Art and Science and the Illustrated Travel Account, 1760–1840* (1984), Hugh Honour described images of America in *The New Golden Land: European Images of America from the Discoveries to the Present Time* (1975), while Bernard Smith prepared a masterly exploration of the interrelationship of science and depiction in his discussion of the impact of scientific discovery on conventions of English landscape painting in *European Vision and the South Pacific* (1988).[13] For all these writers, trained in the analysis of the work of art, the image has value as primary evidence. Art historians certainly ask different questions of the image than do historians of science, but their views on the primacy of the image and the importance of analysing it carefully as an intelligent and intelligible rendering of thought are a useful tonic to historians' arbitrary neglect.

Another group of scholars who have seen the image as more than simply decorative are the historians of printed books and book illustrations. This genre encompasses such works as S. Peter Dance's *The Art of Natural History: Animal Illustrators and Their Work* (1976), A.M. Lysaght's *The Book of Birds*

(1975), Peyton Skipwith's *The Great Bird Illustrators and Their Art, 1730–1930* (1979), and Sacheverell Sitwell, H. Buchanan, and J. Fisher's *Fine Bird Books, 1700–1900* (1953).[14] These illustrated folios, destined primarily for the lay market, and often dismissed as examples of mere connoisseurship, are, however, of great value in a study of representation. Not only do these widely available works offer a wealth of reproductions from both manuscript sources and rare books, but the authors have also provided careful analysis of the connections between the various works – of the borrowings, copyings, and the new interpretations embodied in the images. Finally, there exists a series of reference works devoted to an analysis of the literature of illustrated natural history itself. These include the books of Jean Anker, Canon Raven, Wilfrid Blunt, Agnes Arber, and D.M. Knight. For a variety of reasons these specialist studies have not entered into the historical mainstream and are little known outside the world of natural historians and bibliophiles, though Raven's and Arber's works, in particular, have much to offer concerning the world-view of the eras they discuss.[15]

One historian who has suggested that a re-examination of the role of images as historical evidence was overdue is Francis Haskell, who acknowledges that indeed historians have left the examination of the 'evidence offered by art' to the 'antiquarians.' He describes his 1993 book, *History and Its Images: Art and the Interpretation of the Past*, as being 'concerned with a dialogue ... that has been stimulated by the claims of those who have tried to insist that an image can be seen as a valuable historical source.'[16] Haskell is cautious, however, in seeing the image (or the artifact) as a source for historical analysis, because he notes the problems inherent in the persistence of imagery – the unfortunate fact that printers and authors in the past were by no means scrupulous in their creation and use of images. They appropriated plates from other publishers and re-engraved or redrew familiar images to complement the text, extracting the image from its original context and in some cases even changing the caption to present, for example, a portrait of an old man as a historical personage. In other cases, publishers provided specially engraved images to embellish a historical account, even though it would have been obvious to the readers that the images of ancient kings were of necessity products of an artist's imagination. What is the historian to make of the evidence offered by these ahistorical uses of images in the past, and how much 'evidence' can be derived from their analysis? Haskell is reluctant to view images in isolation and notes that 'what we choose to call art is indeed best interpreted by the historian when it is studied in conjunction with other available testimony ...' While there is no question that the relationship between image and text is highly significant in understanding either, what of the testimony of the raw representation? Haskell suggests that 'it does have a "language" of its own which can be understood by those who seek to fathom its varying purposes, conventions, styles and techniques.'[17]

Few have been better qualified than William Ivins, Jr to read the language of the image. Ivins was for many years Curator of Prints at the Metropolitan Museum of Art in New York, and in the preface to his 1953 classic, *Prints and Visual Communication*, he wrote that the book's thesis 'grew out of a long endeavour to find a pattern of significance in the story of prints.'[18] In the museum world in which Ivins worked, prints were relegated to the status of a minor art form, a lesser vision, inferior to the original drawing or painting. What Ivins realized as he began to search for the pattern of significance is that prints could be defined not only as bearers of aesthetic qualities, but also, from a functional point of view, as bearers of communication. As such, they had a transforming power, and 'far from being minor works of art, prints are among the most important and powerful tools of life and art.' He defined prints as 'exactly repeatable visual or pictorial statements,' and pointed out that, from the middle of the fifteenth century to the mid-nineteenth century, the print was the only means available to convey complex visualizations. Much is made, he notes, of the invention of the printing press in Europe and its transforming effect on social and intellectual life, but the role of the exactly repeated pictorial statement had, by and large, been ignored by historians (as Rudwick subsequently noted). Ivins suggests that, 'if we define prints from the functional point of view ..., it becomes obvious that without prints we should have very few of our modern sciences, technologies, archaeologies, or ethnologies ...'[19] At first reading it might appear that Ivins claims too much for his particular medium, but if, as he insists, we begin to think about the transmission of detailed information on techniques, on the construction of machines, or even on landscapes, or on flora and fauna, we begin to understand the importance of the well-rendered image in communicating knowledge. While Ivins's perception of the functional importance of prints is significant, what is even more germane is his understanding of the limitations that technique imposes on the print as a conveyor of information and on the reader as a receiver of information. Just as the interpenetration of text and image is critical to analysing the nature of representation, so is the relationship between the appearance – a result of a physical process – and content.

Ivins's insistence on the functional role of the printed image and its importance in scientific understanding and change is also reflected in a number of recent publications on the relationship between picturing and information, a relationship which is gaining in significance as visualization becomes a tool of analysis as well as communication. Edward Tufte brings together in *Envisioning Information* (1990) a fascinating series of examples of the problems (and successes) encountered in rendering complex information in graphic form. His examination of the translation of railway timetables, for example, which embody both temporal and spatial information, to the two dimensions of 'flat land' reveals the powerful nature of graphic representation, which makes the complex comprehensible literally at a glance. What is even more significant is that a visualization can also sometimes reveal previously unsuspected relationships between items of information.[20] This is part of the new importance that visualization has achieved in the world of computer graphics and imaging. The editors of a 1993 publication on the use of computer-generated imaging in science, *Focus on Scientific Visualization*, bring together a number of papers that go beyond flat land into the illusory multidimensional world of the computer screen, where the resolution of complex problems in more than three dimensions has become commonplace. Scientists have turned to visualization as a means of presenting complex data, such as fluid flow systems, which they assert can be understood better visually than numerically. While complex mathematical calculations may form the ground of the experiment, their computer-dependent complexity prohibits them from being easily understood. 'A major means to get insight is to visualize the data.'[21] Visualization, then, has assumed new importance in the making of science as well as in its illustration. The functional analysis of imagery changes the way in which the 'visual evidence' can be interpreted. Rather than seeking to understand what the image can tell us as a reflection of the past, we can start to examine the image's function for users in the past. This perspective begins to help us approach the notion of dissonance that we encountered in looking at Edwards's *Bittern*. A principal interest of this book, then, is the relationship between complex information, and the complexity and accuracy of its representation. How important was accuracy in scientific illustration and informative images? What was the information which required visualization in the images of the sixteenth, seventeenth, eighteenth, and early nineteenth centuries? How did naturalists and scientists use the visualizations in their work and in their thought? What role did the scientific image play in envisioning information about the New World for European readers?

Haskell noted the need for understanding the language of art as critical to a reading of visual evidence. Reading images demands a set of tools with which historians are, by and large, unfamiliar. While art historians have certainly developed an enviable set of skills in reading the image as a work of art, these do not sit entirely comfortably on images whose original purpose was primarily informational, as opposed to aesthetic. If we wish to make inferences about the purposes for which images were used by naturalists and scientists and the role they played in the making as well as the transmission of scientific ideas, then we must begin to pay particular attention to the way in which such images were created, reproduced, used, and read. Ivins has already alerted us to thinking about the techniques by which images are made, and the study of their material nature, using the tools of material history, can begin to help us find the language that allows the 'mute' image to speak. Material history is a special branch of the discipline often confined to the museum and relegated by most historians to that curious realm inhabited by the 'antiquarians.' Material historians would suggest, however, that the neglect of non-textual evidence has barred many scholars from a deeper appreciation of the texture of the past. Steven Lubar and David Kingery note in the introduction to *History from Things: Essays on Material Culture*, that, 'by neglecting all but a narrow class of artifacts, those with writing on them, historians have missed opportunities. Artifacts are remnants of the environment of earlier periods, a portion of the historical experience available for direct observation.'[22] On the basis of this direct observation, material historians aspire to create a history of mind, an attempt to understand the *mentalité* of a group of people, revealed in the objects they create.[23] Material historians tend to make a distinction between works of art and artefacts, not simply as the result of institutional organization that divides the art gallery from the museum, but because, as Jules Prown points out, 'artifacts do not lie.'[24] What he is referring to is the primarily utilitarian function of the artefact. While works of art are inevitably self-conscious, most artefacts are not. Rather, they reflect the 'underlying cultural assumptions and beliefs' that are revealed in the style of the object.[25]

The images of natural history and landscape I have chosen to examine are not usually categorized as 'works of art.' In some cases they are anonymous; in others, a collaboration of a number of hands, passing from sketch to engraving to print. Their primary purpose is not to evoke aesthetic response, but rather to communicate information, yet inherent in each image is a reflection of the style of its day. Each embodies certain assumptions about the makers and the viewers of the object, their interests and their understandings, the cultural lens through which they view the image. While I have not limited myself to the evidence of image alone, the adoption of a material historian's perspective ensures that the image is treated as primary rather than supplementary, and that the functional aspect so significant for Ivins is highlighted. Moreover, a material-history analysis is accomplished on the object's own terms. Such an analysis consists of a complete description of the object, including the material from which it is made, the manner of its making, its age, its culture, and, where possible, a description of its maker. Material-history analysis often relates the object to similar objects from the same culture or from others. It sometimes goes further to speculate upon the purpose for which the object is made and to discuss its iconography, the internal coherence of its decoration or depiction. In many ways, the practice of the material historian embodies the precepts enunciated by Michel Foucault for the 'analysis of the discursive field' described in *The Archaeology of Knowledge*. Foucault suggests that the history of thought, of ideas, of science, has been too concerned with allegorical analysis – what was really being said? – and the search for origins – which was the original statement, which the banal repetition? Foucault suggests that a more fruitful approach to the analysis of discourse (in which we can include images, as Foucault himself does) is the analysis of specificity: 'we must grasp the statement in the exact specificity of its occurrence, determine its conditions of existence, fix at least its limits, establish its correlations with other statements that may be connected with it, and show what other forms of statement it excludes.'[26] He describes this type of investigation as 'archaeological,' and it must be assumed that Foucault has not chosen his metaphors lightly. By looking closely at things, archaeologists, like material historians, are able to make statements about the past. For Foucault, discourse can also be embodied in pigment and canvas. 'Archaeological analysis would not set out to show that the painting is a certain way of "meaning" or "saying" that is peculiar in that it dispenses with words. It would try to show that, at least in one of its dimensions, it is discursive practice that is embodied in techniques and effects ... It is shot through – and independently of scientific knowledge (*connaissance*) and philosophical themes – with the positivity of a knowledge (*savoir*).'[27]

It is between Foucault's notion of *savoir* and the material historian's desire to chart the track of mind, or what the American historian Henry Glassie calls 'the architecture of past thought,'[28] that this analysis falls. The discursive archaeology that Foucault proposes is perhaps closer to material history than to traditional document-based historical analysis. By looking at sixteenth- and seventeenth-century images of the New World, the contemporary reader is not trying to see how North America looked in the past, but how it was depicted by those who created and reproduced the image. Stephen Greenblatt, in his analysis of Columbus's discourse, has adopted Foucault's approach to these late-fifteenth-century words and images. He writes: 'I have tried less to distinguish between true and false representations than to look attentively at the nature of the representational practices that the Europeans carried with them to America and deployed when they tried to describe to their fellow countrymen what they saw and did.'[29] Similarly in this analysis of the discourse represented in images of natural history, I seek to understand each image in its specific creation, to describe, as Foucault and Greenblatt suggest, the discontinuities, the rifts, the particulars embedded in and surrounding each thing. By examining this array of documents of a different order, different conclusions may in fact be drawn about the understanding of nature and the world in the past. David Knight has pointed out in his book *Zoological Illustration: An Essay towards a History of Printed Zoological Pictures*, that, in looking at scientific illustration, we may discover that our notions of science are culturally mediated: 'We do not therefore simply find progress in zoological illustrations ... Indeed to approach the history of zoology through illustration, is a ready way of dropping the idea of science as cumulative progress to indubitable truth based upon some "scientific method."'[30]

In this book, then, I undertake at one level a material-history analysis of a series of images arranged in roughly chronological order. The objects I have chosen for this analysis are almost all two-dimensional works on paper, canvas, vellum, and so on. All the objects include depictions of plants or animals, and sometimes I treat the image as a whole, and sometimes the component parts, in an iconographical discussion. This material-history analysis is imbued with the strictures that Foucault has laid out concerning 'archaeology.' Foucault recognizes that change as the engine of history is problematic. He insists that archaeological analysis 'considers [that] the same, the repetitive and the uninterrupted are no less problematic than the ruptures.'[31] I am not interested in showing how artists learned to draw better bears or bison or birds, but how they used images to communicate certain understandings about the natural world, how their very depiction of the real world (and what can be more real than a particular bird or a flower?) reflected their interests and their knowledge. From the sixteenth century onwards, artists insisted that their images were 'drawn from nature,' 'true and accurate descriptions,' or 'drawn from life.' It is obvious even from a cursory comparison of images from Thevet's *Cosmographie universelle* (1575) or Topsell's *History of Four-Footed Beasts and Serpents and Insects* (1658) with those from Rondelet's *Libri de Piscibus Marinis* (1554) or Linnaeus's *Hortus Cliffortianus* (1737 [1738]) that 'drawn from life' had very different meanings for different authors and artists. By looking attentively at what Rudwick referred to as 'the strong visual component of the original source-materials,' I hope to be able to make some contribution to an understanding of the transformation of the world of Columbus and Cartier into that of Darwin and Humboldt, and essentially into our own.

The images have been selected according to a set of criteria which developed over the course of my examination of a large number of original maps, prints, and drawings in collections in North America and Europe. All images contain representations of the flora, fauna, or natural landscape of the New World, and all were developed in the European tradition. While I attempted to confine my selection to those images with 'Canadian' content, I have of necessity had to broaden the range, though I have tried in almost all cases to deal only with the Nearctic as opposed to the tropical regions. Many early works, such as those by White and Catesby, though they deal with the fauna and flora of what are now the southern United States, were used and copied by those who explored and wrote about more northern lands. In most cases, I have also selected only those images which had a wide distribution, either in print or through circulation among a group of specialists. Thus, I have included Georg Ehret's watercolours of Newfoundland plants, not only for their delicate beauty, but also because as part of Joseph Banks's collection they were available for consultation by Banks's great and influential circle of like-minded friends and acquaintances. I have also included the watercolour sketches of George Back and Robert Hood from the Franklin Overland Expedition since these were later published as engravings to illustrate the official account. I have not, however, included, despite being urged to, the manuscript drawings of the celebrated *Codex Canadiensis*, which, though undeniably interesting, were never published until released in facsimile in 1930, and most recently in 1974 by Éditions du Bouton d'Or, Montreal, as *Les Raretés des Indes: 'Codex Canadiensis.'*

The first series of images I examine are embedded in the early-sixteenth-century maps of the northern part of the New World. While most attention was turned to what is now South America, cartographers did incorporate images of the birds, beasts, and flora of the north into their works. I also look at concomitant images of New World animals as they appeared in very early travel books and records of voyages. The second set of images centres around botanical illustration. Scholars have always remarked on the exactness of botanical as opposed to zoological illustration in works of the sixteenth, seventeenth, and early eighteenth centuries. Certainly the flora of the New World is thought to have had a profound influence on the development of systematic botany in the eighteenth century. Third, I look at the flood of images of flowers, birds, beasts, and fish that emerged from the mid- to late-eighteenth-century European presses, the age of George Edwards and his colleagues. This eighteenth-century tradition is followed in another group of images which sees the culmination of the conventions of natural-history observation and illustration in the early nineteenth century. The concluding chapter attempts to draw together the threads of the analysis in a weave which examines the role of image in the understanding of nature and the logic of science.

one

Emblematic Animals

In 1492, Christopher Columbus (1451–1506) set the standard of Spain on his first landfall in the New World. When he returned, he wrote a letter to Luis de Santangel, who had helped him to find financing for the voyage, describing the journey and the lands that they had seen. The 'Columbus letter' was first published in Barcelona as a broadsheet in March or April 1493, and over the next five years was republished in at least seventeen editions, including a Latin translation, making it available to erudite readers throughout Europe. It had also been translated into sixty-eight stanzas of Italian verse by Giuliano Dati, 'meant to be sung in the streets.'[1] Both the Latin version published at Basel in 1493 and the Dati poem were illustrated. The frontispiece of the Basel edition shows the *Insula hyspana*, on which two groups of naked Native people, some with a good deal of trepidation, meet Columbus, who arrives in the ship's boat, bearing in his hands a jewelled cup. It is thought that the publisher used woodcuts adapted with slight modifications from an earlier book depicting Mediterranean ports, which might explain the presence of the trireme, a vessel suited more to the protected waters of an inland sea than to open ocean.[2] Hugh Honour suggests that the Dati illustrations, on the other hand, were specially commissioned. The frontispiece in the edition of the poem shows a seated king, and ships approaching an island, with a palm tree in the foreground and long-haired Native people, the women wearing girdles of leaves or feathers.[3] Certainly Columbus mentioned palm trees specifically in his letter ('There are palm trees of six or eight kinds, which are a wonder to behold because of their beautiful variety'), and the fact that some of the Native people 'have the custom of wearing their hair long like women.'[4]

But neither of the frontispieces conveys the landscape so well as Columbus's written description of Cuba:

> Its lands are lofty and in it there are many sierras and high mountains ... All are most beautiful, of a thousand shapes, and all accessible and filled with trees of a thousand kinds and tall, and they seem to touch the sky; and I am told that they never lose their leaves, which I can believe, for I saw them as green and beautiful as they are in Spain in May, for some of them were flowering, some with fruit and some in another condition, according to their nature. And there were singing the nightingale and other little birds of a thousand kinds in the month of November, there where I went. There are palm trees of six or eight kinds ...; there are marvellous pine groves, and broad fertile plains, and there is honey. There are many kinds of birds and varieties of fruit.[5]

In his diary of the first voyage, Columbus also notes the fish, 'so different from ours ... of the finest colours of the world: blues, yellows, reds, and of all colours; and others coloured in a thousand ways,' and the 'flocks of parrots that obscure the sun; and birds of so many kinds and sizes and so different from ours ...'[6] In these passages imagination mediates between the written description and the mental image, producing vivid representations of an exotic world in the minds of the readers, whose exposure to tropical flora and fauna was conditioned by geographic location and by exposure through iconic images and imported specimens. The pictorial representations do not, however, realize the complexity of the text, conveying an idea (the idea of arrival, the idea of Native people) rather than a physical reality.

Columbus's initial voyage was followed by three additional voyages under his command, and his example served as inspiration not just for the Spanish court, but for the Portuguese and the English, and eventually the French. In 1497, John Cabot sailed to Newfoundland under the English flag, to be followed in 1500 by Gaspar Corte-Real, who coasted Labrador. Amerigo Vespucci sailed for Brazil in 1499, and has been credited with the first pronouncement that the lands of the western sea were 'what we may rightly call a New World ... a continent ...'[7] In his published letters of 1505 and 1509, Vespucci describes the Native people with their feather ornaments, as well as the brilliant flowers, strange birds, and even a dragon (most likely an iguana).

Unlike Columbus and Vespucci, who quickly published accounts of their voyage, neither Cabot nor members of the Corte-Real voyage wrote detailed records of their explorations. Certainly accounts of their voyages were reported to the authorities, and the Native people and strange animals they brought back with them displayed to kings and commoners, whose reactions were recorded. Alberto Cantino, the envoy of the Duke of Ferrara, for example, notes that he had 'seen, touched and examined these people'[8] kidnapped by the Portuguese expedition, and Henry VII paid £5 'to Portyngales that brought popyngais & catts of the mountayne with other stuf to the King's grace ... of the New-found Island.'[9] Cantino also commissioned a map for his employer to document the discoveries of the Portuguese and Spanish upon which the Italians were casting an apprehensive eye. In a 1501 letter to the Duke, Cantino recounts Corte-Real's voyage to Newfoundland: 'they found abundance of the most luscious and varied fruits, and trees and pines of such measureless height and girth, that they would be too big as a mast for the largest ship that sails the sea.' In another letter, Cantino, whose descriptions embody the feeling of first-hand accounts, relates that 'no corn grows there, but the people of the country say they live altogether by fishing and hunting animals, in which the land abounds, such as very large deer, covered with extremely long hair, ... and again wolves, foxes, tigers, and sables. They affirm that there are, what appears to me wonderful, as many falcons as there are sparrows in our country ...'[10] The Cantino map thus shows not only the parrot-infested South American coast, but also the *Terra del Rey de portuguall* covered in tall trees.

Despite the Native people and the 'blue popyngais,' the cold north did not touch the European imagination in the same manner as the warm and exotic south, and by very early in the sixteenth century the dominant image of the New World was grounded on the experience of the Indies and of Central and South America. As Samuel Eliot Morison points out, after the Spanish and Portuguese voyages to India, Brazil, and Venezuela, 'who cared for codfish, mast trees and icebergs?'[11] In 1520, the treasures of Montezuma were on display in Europe, and that glittering horde fixed the image of America as strange, rich, and exotic, full of peculiar beasts and wonderful plants. By the mid-sixteenth century, America was personified, in countless allegorical drawings, engravings, and paintings, as a young, naked woman, sometimes with feathered headdress or girdle and arm- and legbands, accompanied by a parrot, and often associated with the remains of cannibal feasts – severed heads and limbs, torsos on a spit, body parts in baskets – sitting on that strange New World beast, an armadillo, or perhaps an alligator, or in one case reclining in a hammock. The image of America as tropical led, however, to confusion with the tropical kingdoms of the Old World. According to Morison, Vasco da Gama's voyage to Calcutta in 1497–9 attracted more attention than Columbus's Atlantic crossing. A procession in honour of Maximilian in 1516 shows a group of people clad in feathers, bearing parrots, prehensile-tailed monkeys, tropical fruits, and sheaves of corn – the products of the New World. They are called, however, the 'people of Calicut' and are accompanied by fat-tailed Asiatic sheep that appear in many allegories and prints of America. The tendency to confuse Asiatic India and the Indies of the New World continues throughout the sixteenth century, resulting in the misappellation of the turkey, and even of corn, commonly called 'turkey wheat,' as if all that was new and strange must hail, not from across the relatively unknown western ocean, but from the familiar if distant East.

One reason for this lack of interest in the northern hemisphere can be explained by the almost complete absence of the exotic. As Carl O. Sauer has observed of the northern voyages, 'nowhere else could they have gone so far across the sea and found so much so familiar to them.'[12] While some of the flora and fauna were definitely strange, much was common to both sides of the Atlantic, from the beavers, bears, and falcons to the oaks, firs, and beeches. In his first voyage of 1534, Jacques Cartier (1491–1557) describes the coast of Prince Edward Island:

> We landed that day in four places to see the trees which are wonderfully beautiful and very fragrant. We discovered that there were cedars, yew-trees, pines, white elms, ash trees, willows, and others, many of them unknown to us and all trees without fruit. The soil where there are no trees is also very rich and is covered with pease, white and red gooseberry bushes, strawberries, raspberries, and wild oats like rye, which one would say had been sown there and tilled. It is the best-tempered region one can possibly see and the heat is considerable. There are many turtle-doves, wood-pigeons and other birds. Nothing is wanting but harbours.[13]

On the third colonizing voyage of 1542, he describes the land near Cape Rouge, Quebec:

> More[o]ver there are great store of Okes the most excellent that ever I saw in my life, which were so laden with Mast that they cracked againe: besides this there are fairer Arables, Cedars, Beeches, and other trees, then grow in France: and hard unto this wood on the South side the ground is all covered with Vines, which we found laden with grapes as blacke as Mulberies, but they be not so kind as those of France because the Vines bee not tilled, and because they grow of their owne accord. More[o]ver there are many white Thornes, which beare leaves as bigge as oken leaves, and fruit like unto Medlers. To bee short, it is as good a Countrey to plow and mannure as a man should find or desire. We sowed seedes here of our Countrey, as cabages, Naveaus, Lettises, and others, which grew and sprong up out of the ground in eight dayes.[14]

In his accounts Cartier is able to name many of the plants he sees. He even recognizes that, with cultivation, the wild productions of this new world could be made to resemble those of France, and, in fact, the soil seems to be suited to the produce of France, bringing seedlings out of the ground in eight days, an assertion he repeats twice. Not only is the vegetation similar, but Cartier can name a dozen birds, including 'cranes, swans, bustards, geese, ducks, larks, pheasants, partridges, blackbirds, thrushes, turtledoves, goldfinches, canaries, linnets, nightingales, sparrows and other birds, the same as in France, and in great numbers.'[15] Columbus, on the other hand, cannot name what he sees. In his diary he writes of an island of 'many very green and very large trees ... I do not know where to go first; nor do my eyes grow tired of seeing such beautiful verdure and so different from ours.' The birds, too, are 'of so many kinds and sizes, and so different from ours, that it is a marvel.'[16] Even the fish of the Indies are of a thousand colours, unrecognizable, not like the cod that throng the northern waters.

Cartier's narratives of his voyages were not published until the mid-sixteenth century. That for the second voyage was printed in Paris in 1545, while the record of both early voyages was included in Giovanni Battista Ramusio's compilation of 1556, *Navigationi et Viaggi*, which contained an illustrated map by Giacomo Gastaldi (*ca* 1500–*ca* 1565). Cartier's discoveries and those of Cabot were, however, incorporated on manuscript maps produced by the Dieppe hydrographers. These cartographers would have obtained information practically from dockside, and a series of charts dated between 1544 and 1553 and attributed to Pierre Desceliers (1487–1553) of Arques reflects Cartier's discoveries. It was primarily from these map sources and the written narratives that accompanied them that Europeans prior to the mid-sixteenth century drew their images of the flora and fauna of the northern half of the New World. The noted cartographer R.A. Skelton recognizes the significance of the map as a visual representation in the development of geographic understanding. It is, he writes, 'a graphic document employing visual symbols,' and as such 'it has made a more immediate impact on men's minds than the written word. This may be partly explained by the positiveness with which it presents geographical facts. The statements which it makes – on position, direction, distance, or extent – have an absolute character, in comparison with those written texts which can more easily be qualified.'[17] Early sixteenth- and seventeenth-century charts were not only used by sailors and travellers; the highly ornamented manuscript charts were 'made by professional artist-cartographers not for the purposes of navigation, but to adorn the salons and studies of kings, cardinals and wealthy patrons of geography.'[18] In the absence of other visual information, the maps often afforded those interested in the recent discoveries their first glimpse of a new world.

Glimpses of the New World: Early Maps of North America

The world map of 1546, now preserved in the John Rylands Library in Manchester, has been attributed to Pierre Desceliers, one of the great cartographers of Dieppe. It is based on information derived from the voyages of both Cartier and Roberval, and even contains a portrait of 'monsr. de Roberval' addressing his men at 'Le Sagnay.' The map is highly coloured, and the original was painted on parchment for Henri II. The illustration (plate 2; see colour plates) is taken from a nineteenth-century redrawing of the original, lithographed by E. Rembielinski and published at Paris by E.F. Jomard in *Monuments de la géographie*. Contrary to current conventions, north is at the bottom of the map, and the image appears to present-day viewers to be upside down. The northern seas, the 'Mer Despaigne' and the 'Mer de France,' which surround the continental mass, are populated with whales and strange fish. One vignette depicts Native hunters harpooning a whale that sports long whiskers (baleen?).[19] In the 'Mer des ĝtilles,' there are two creatures with horse-like heads and long teeth, which may be the sea morse or horse, the walrus. On land, in 'La Terre du Laboureur,' or Labrador, two knights meet in joust while men-at-arms watch. Here, the land is hilly with few trees, but overall the northern land is mountainous, with many tall trees, as in Cartier's reports. In 'Canada,' on the north shore of the St Lawrence, are two large deer, two bear, and a large dark bird which might be an eagle or bustard, or the large dark falcon described by Sebastian Cabot ('there are also in this country dark-coloured falcons like crows, eagles, partridges, sandpipers, and many other birds').[20] In the 'campestra bergi' to the west of 'Ochelaga' are found a peccary, or wild boar, and a porcupine. Another peccary is being hunted by a Native person on the east coast. The east coast midway between Florida and the St Lawrence also sports a unicorn, while a white horse is found inland. Farther south, towards 'Nueve Espaigne,' is a grouping of five animals. One resembles a pale-coloured wolf or coyote; the two brownish animals are perhaps coyotes, or even foxes; while the other pale animal with a long tail appears more like a cat (perhaps a cougar?). The strange, horse-shaped animal with the long tail may be an attempt to represent the bison[21] or perhaps the tiger or jaguar. The unicorn is the only truly exotic creature, and perhaps Desceliers has included it in the spirit in which, two hundred years later, Michael Collinson wrote to the American naturalist John Bartram: 'With regard to the unicorn I am rather divided in my judgment, even in respect to their present existence, in the interior region of Africa, of which at this period we are extremely ignorant.'[22] Since the unicorn no longer appeared to exist in Europe, perhaps it existed in this other temperate continent, of which so much was yet unknown. Wilma George notes, however, that before mid-century 'bears, coyotes and deer occur more frequently than any other animals on North American maps.'[23] Bears appear on a 1519 manuscript map along with deer and a coyote(?), and on Sebastian Cabot's 1544 engraved world map. The first polar bear, however, does not appear until Desceliers' 1550 map, where it is devouring fish on ice floes off the coast of 'Terre du Labrador.' Cartier had in fact reported on his first voyage that his men found a bear on an island offshore 'as big as a calf and as white as a swan that sprang into the sea in front of them.'[24]

PLATE 3: Giacomo Gastaldi (*ca* 1500–*ca* 1565), ***Map of New France,*** 1556. Woodcut. From Giovanni Battista Ramusio, *Navigationi et Viaggi* (Venice, 1556). This is the first of the map's two known states.

While Desceliers' manuscript maps reported the discoveries of the French explorers with a certain degree of accuracy, the first published map of New France, an illustration that accompanied the account of Cartier's 1534 voyage in Ramusio's *Navigationi et Viaggi,* did not, in fact, reflect the details related by Cartier but rather those of the 1524 voyage of Verrazzano. It is, however, well populated with plants, animals, and Native people, and depicts in schematic format the Grand Banks. The map reproduced here (plate 3) is the first of its two known states. It is a woodblock print by the Venetian cartographer Giacomo Gastaldi, and shows 'La Nuova Francia,' which includes 'Terra de Nurumbega,' and a group of islands identified as 'Terra Nuova' and 'Isola de Demoni.' Again the seas are inhabited by fearsome monsters, one resembling a dog, or perhaps a sea wolf, others spouting like whales. At the same time there are several vignettes of men fishing from small boats all around the islands of 'Terra Nuova,' some with nets and others hand-lining cod. The Grand Banks are represented by a ribbon which begins in the eastern '(Levante)' portion of the map and ends near the mainland at a head called 'Angoulesme,' which Ganong identifies with New York.[25] The islands of Terra Nuova are primarily the domain of birds, though one bear is shown, and there are also birds in the water between the islands. Cartier describes the bird islands off Newfoundland and is obviously impressed by the number of gannets, murres, puffins, and auks. He also describes the 'tinkers,' small birds that fly in the air and swim in the sea.[26] On the 'Isola de Demoni,' a Native person shoots a bird with outstretched wings resembling an eagle or perhaps meant to represent the dark-coloured (peregrine) falcons described by Sebastian Cabot on the 1544 world map. On this northernmost island are also pictured a group of Native people under a roofed construction made of poles, similar to shelters on the mainland described briefly by Verrazzano. Strangest of all, however, is the depiction of three winged 'demons' with horns and short tails who appear on the land and in the air above the 'Isle of Demons.' There is no mention of these strange creatures, or anything like them, in Cartier's accounts. There is, however, a reference in a letter from one Lagarto to John III, King of Portugal, written perhaps in early 1539, which recounts a series of conversations between Lagarto and Cartier, including one in which Cartier is 'greatly praising the rich novelty of the land and telling these and other tales; and that there are men who fly, having wings on their arms like bats, although they fly but little, from the ground to a tree, and from tree to tree to the ground.'[27] (The description is suspiciously reminiscent of flying squirrels.) On the mainland, Gastaldi shows Native activities, chiefly hunting, perhaps dancing, and conversing. Several Native people are shown using bows and arrows to hunt birds and bears, and perhaps a wild pig. One Native person is roasting a pig(?), and two others are carrying home what looks to be a deer or another bear. Two fish are drying on a pole strung between two trees, and there is a depiction near the coast of what might be a weir. For the first time on any map of North America, a running rabbit or hare appears.[28] On his second voyage, Cartier discovers an island where hares are so plentiful that he names it 'Hare Island.'[29] Finally, by an inland river, a Native person lies under an upturned canoe, something referred to by Cartier in his account of the first voyage.[30]

What is perhaps most surprising on these two maps is not the animals that are depicted, but those that are not. Cartier describes several times and in detail a number of marine mammals, including walrus, seals, and beluga whales. On the first voyage he writes that on an island near Newfoundland they came upon 'many great beasts, like large oxen, which have two tusks in their jaw like elephant's tusks and swim about in the water.' On the second voyage he also saw 'several fish in appearance like horses which go on land at night but in the daytime remain in the water ...'[31] It is on the second voyage as well that he discovers 'a species of fish, which none of us had ever seen or heard of ...': 'This fish is as large as a porpoise but has no fin. It is very similar to a greyhound about the body and head and is as white as snow, without a spot upon it. Of these there are a very large number in this river, living between the salt and the fresh water. The people of the country call them *Adhothuys* and told us they are very good to eat. They also informed us that these fish are found nowhere else in all this river and country except at this spot.'[32]

Cartier, of course, has encountered the beluga whale, once common along the Atlantic coasts and on the St Lawrence River, with a small stationary population still present at the mouth of the Saguenay River, as well as walruses and seals. The cartographers did not, however, attempt to depict these characteristic beasts (unless the horse-headed and tusked sea monsters drawn by Desceliers are indeed seals and walruses), and probably with some justification. Even so great an artist and so sensitive an observer of the natural world as Albrecht Dürer had some problem in drawing from life the walrus. Dürer went to Zeeland in 1520 in the hopes of seeing a stranded whale, but instead found a walrus, of which he did a drawing in pen and brown ink and tinted with watercolours. The inscription reads '1521 / The animal whose head I have drawn here was taken in the Netherlandish Sea and was twelve Brabant ells long and had four feet.' The animal does not quite come off to contemporary eyes as a walrus, and it is possible that Dürer may not have seen a living animal but rather a mounted or dried specimen. Dürer later used the study for the head of a dragon.[33] Over one hundred years later, Johannes de Laët does manage to include a reasonably recognizable engraving (plate 4) of a walrus and calf in his 1633 book, *Novus orbis, seu Descriptiones Indiae Occidentales, libri XVIII*. The engraving, he writes, has been drawn accurately to the life ('*Iconem hic subjicimus ad vivum accurate expressam*'), and he notes details of its tusks, the distinguishing feature of the Walrus, or morsh, as he refers to the animal.[34] Similarly, the cartographers do not attempt the beluga, relying

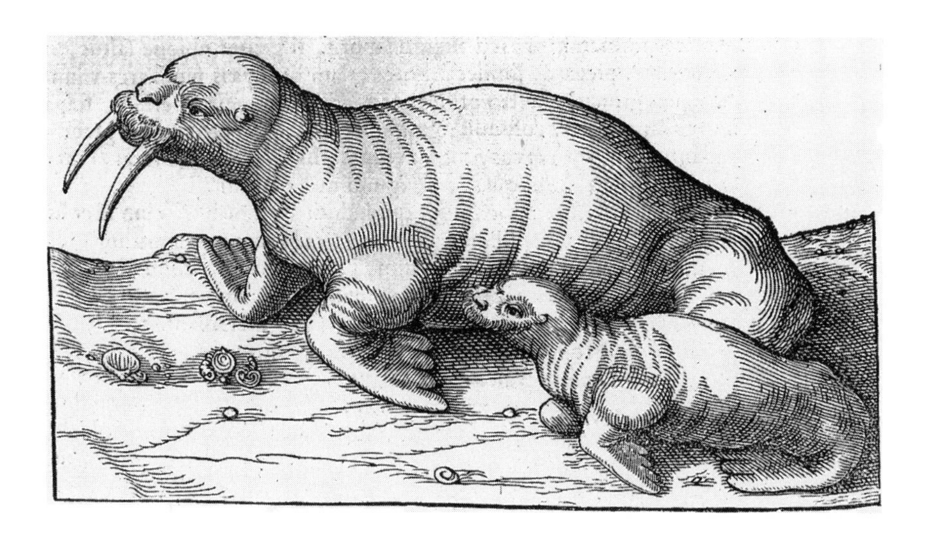

PLATE 4: Walrus and Calf, 1633. Engraving. From Johannes de Laët (1593–1649), *Novus orbis, seu Descriptiones Indiae Occidentales, libri XVIII* (Leyden, 1633), 38. De Laët notes that the walrus has been drawn after life, and it is likely that he saw a preserved specimen in a cabinet. The same illustration is used in the catalogue for Ole Worm's *Museum Wormianum*, published in 1655.

instead on the host of whales and marine monsters provided in the woodcut of sea monsters inhabiting the northern oceans in Sebastian Münster's *Cosmographia*, published in Basel in 1550 and repeated in Konrad Gesner's (1516–1565) *Icones Animalium* in 1560. Similar monsters were featured on Olaus Magnus's 1539 map of the north. Why are these characteristic animals not included by the cartographers, or, if included, rendered in such a bizarre manner? What were the problems inherent in depicting new life forms and why were the same images repeated over and over in maps and books?

Marks and Emblems: Claiming a New World

Jonathan Swift, in a seventeenth-century satiric poem, suggested that

> So Geographers in Afric-Maps
> With Savage-Pictures fill their Gaps;
> And o'er unhabitable Downs
> Place Elephants for want of Towns.[35]

The noted cartographic historian Wilma George is the authority for the depiction of animals on maps. Early voyagers, as noted above, did bring back specimens of the flora and fauna of the lands they visited, and cartographers included these on maps, not, George insists, simply for want of other information. She suggests that, given the strangeness of the flora and fauna, it is not 'surprising that some of the cartographers, following the journals of expeditions or making their own observations on the spot, should have included as part of the land's features some of its peculiar animals, some of its plants. They were decorative but they were in all probability used diagnostically of the countries they inhabited, just as banners identified knights. If there were no towns to put on maps, it is arguable that the animals or plants were as reputable an indication of the terrain as a range of hills or a river.' George goes on to argue that the distribution of the animals on world maps was not arbitrary, not 'a simple desire to picture newly discovered animals,' but rather an attempt to use characteristic animals to symbolize the continents.[36] She also notes that, in terms of recognizing and depicting new fauna, the cartographers were far ahead of the naturalists. She can identify nine new animals on maps by the mid-sixteenth century and only two wholly new animals in Konrad Gesner's celebrated *Historia Animalium* (1551–8). The convention of depicting animals on maps, and their subsequent incorporation into map cartouches, lasts until the late seventeenth century. Two questions come to mind in reviewing the maps and George's thesis. Why did animals find such a prominent place on maps in this period, and why animals and not plants?

The answers to the two questions are related. George notes that cartographers preferred to use animals rather than plants to symbolize the continents because, 'with outstanding exceptions, such as the cactuses, plants do not show clear-cut localisation in the world. In contrast, the localisation of animals was noticeable to the early explorers and clearly of interest to the mapmakers.'[37] But certainly plants were of great interest in the fifteenth and sixteenth centuries, particularly as regards their medicinal use. Cartier, in his *List of Men and Effects for Canada* dated September 1538, noted: 'Item, two Apothecaries, each with an assistant, to identify plants and determine their uses ...'[38] Cartier also imported plants, as well as Native people and animals, and the thuja (northern white cedar, or arbor vitae) in the Jardin Royal at Paris came from his voyages.[39] The depiction of plants in printed books, particularly herbals, was also well advanced in the early sixteenth century, especially after the publication in 1530 of Otto Brunfels's *Herbarium Vivae Eicones* with woodcut illustrations by Hans Weiditz (discussed in chapter 3). Yet on the maps there are only generic 'trees,' vaguely deciduous, despite Cartier's descriptions of a wide variety of trees, bushes, and grasses. And while it is true that the plants are not localized, neither can that be said for the animals that roam the North American cartographic world.

The fauna from the two maps includes deer, bear, porcupine, peccaries, wolves, foxes, rabbits or hares, cougar (tiger), seabirds, and eagles or falcons, along with sea monsters and fish (cod). The Nearctic shares a large number of these with the Palaearctic, with the exception of the tree-porcupine and the cougar. While early explorers might marvel at white bears and shaggy deer, they were not unknown to travellers in the North. Olaus Magnus's 1539 map includes a frieze of characteristic northern animals, including reindeer (caribou), foxes, gluttons (identified with the wolverine), and spotted lynx.[40] Animals were, however, chosen as 'diagnostic,' in Wilma George's word, because they identified lands, just the way banners identified knights. It is important to note, however, and this idea will be explored later, that the diagnostic animals are not necessarily very precisely rendered. They are cookie-cutter bears, deer, and rabbits, seen in profile. George has suggested that they were like banners, and there is in that observation a clue to the use of animals on maps. There is a tradition of animal illustration that first found expression in the bestiaries of the Middle Ages and was continued in the emblem-books of the early modern period. The medieval bestiary was a compendium of animal fables and stories descended in part from the Latin *Physiologus* with additions from Isidore of Seville, Gerald of Wales, and other compilers. The bestiary was not valued as a descriptive natural history, but rather as a prescriptive guide; its authors were concerned, not with facts, but with the moral and doctrinal implications of the animal tales they included. The British historian of the early naturalists, Canon Raven, makes it clear, however, that the bestiary tradition did not disappear under the weight of the new scientific interest in natural history that began with the Renaissance, but co-existed with it:

> ... it is well to remember that during the whole period with which we are dealing the medieval outlook was still the popular Weltanschauung, that

> scientific study and the inductive method were hardly appreciated at all until the middle of the seventeenth century, and that the outlook of Bartholomew even in his most extravagant beliefs was still expounded and published not only by poets and preachers and the writers of emblem-books, but as authentic natural history.[41]

The *De proprietatibus rerum* of Bartholomew the Englishman, to which Raven refers, was written in the twelfth or thirteenth century, yet remained the standard authority on natural history up to Shakespeare's time. Emblem-books, too, were a characteristic expression of the Renaissance. The first was published in 1531 by an Italian lawyer, Andrea Alciati, and in these illustrated books were included 'the traditional images, the Pelican in its Piety, the Stork or Crane keeping vigil with a stone in its claw, the Salamander in the Flames, the Lily and the Palm,' with their traditional meanings. The emblems were symbols, available to be used in art or decoration, but with the resonance of nearly seven hundred years of meaning. Francis Quarles noted in his 1635 book *Emblemes*, 'By the knowledge of letters God was known by Hieroglyphics, and indeed what are the heavens, the earth, nay every creature, but Hieroglyphics and emblems of his glory?'[42]

The animals of the bestiaries and the emblem-books stood not just for themselves, but as signs or marks of another reality. As such they did not have to be 'true to nature'; rather, they were true to the symbolic idea of the animal. Thus, on the maps appear bear, deer, and fox, familiar, recognizable animals now used in unfamiliar lands to mark the territory as part of God's creation. Foucault has discussed the concept of the 'mark' or sign and its relation to language in *The Order of Things*. His archaeological analysis of the discourse of the fifteenth- and sixteenth-century naturalists and philosophers reveals a profoundly different way of knowing and a different content to the known. The natural world was a text to be read to ascertain the nature of God. Everything was relevant, and the relations were understood in terms of sympathy or resemblance. This general episteme resolved in the seventeenth century into the well-publicized doctrine of 'signatures,' or signs, which can be revealed, as Paracelsus says, to the person who seeks: 'It is not God's will that what he creates for man's benefit and what he has given us should remain hidden ... And even though he has hidden certain things, he has allowed nothing to remain without exterior and visible signs in the form of special marks – just as a man who has buried a hoard of treasure marks the spot that he may find it again.'[43]

The idea of the mark informed the understanding not just of philosophers and naturalists, but of the sailors and explorers who saw with their own eyes the marks left by God in the New World. In *La Cosmographie universelle* (1575), André Thevet recounts the story of the discovery of maple sap by Cartier's men:

> Among others there is one called *Cotony* ... This tree was [considered] for a long time as useless and without profit until one of our people wanted to cut one. As soon as it was cut to the quick a liquor came out of it in quantity. This being tasted, was found to be of such good taste that some thought it to be equal to the goodness of wine ... And to see and experiment on the source of this drink, the said tree was sawed down and its trunk being on the ground a miraculous thing was discovered in the heart of the tree: a Fleur-de-lys well pictured, which was admired by all. About this some said that it was a very good presage for the French nation, which in the passage of time through the diligence and zeal of our kings could conquer and some day bring to Christianity this poor barbarous people.[44]

Animals were chosen, then, to be banners or emblems of the new lands because they already had significance as symbols in the popular imagination. At the same time, however, that the emblem-books and bestiaries still provided a system of signs and marks by which the world could be understood, a newer, more modern and more scientific viewpoint was emerging. It is perhaps this viewpoint that made Desceliers picture a white bear and a turkey on his 1550 map along with the usual fauna. As the explorers and fishermen returned with tales of the new things they had seen and the strange, if generally recognizable animals they encountered, the map-makers began to respond. As Wilma George points out, they responded faster than the naturalists. Illustrated natural histories such as Gesner's *Historia Animalium* were an innovation, and preparing the illustrations was not an easy task. It was made even more difficult when artists and illustrators were faced with the depiction of the wonders described in travellers' tales. How to draw what they had never seen and could scarcely imagine? They sought inspiration not only in the text, but also in the melding of the familiar with the strange to create the exotic.

Words into Pictures: The Su, the Bison, and the Simivulpa

While the map fauna of the early sixteenth century depicted bears, deer, and foxes, Cartier had also spoken of animals new to the explorers: walruses, 'like large oxen, which have two tusks in their jaw like elephant's tusks,' and beluga whales, 'as white as snow and have a head like a greyhound's.' He also described muskrats, 'which live in the water and are as large as rabbits and wonderfully good to eat.'[45] These animals did not appear on maps, although, by the mid-seventeenth century, beavers and otters were included on Nearctic maps. Why did the artist-cartographers not attempt to include the peculiar fauna of the northern hemisphere on the map, as they had that of the southern hemisphere (macaws, armadillos, llamas, and capybaras)? Part of the answer lies in the lack of models from which to draw, and part in the conventions of illustration.

Brian Ford points out that 'most scientific illustrations are drawn from earlier reference images, rather than from life. The creation of "icons" – unrealistic images that pass down from one generation to the next – is a feature of science texts.'[46] Simply put, it was easier to copy the image of an animal already rendered into two dimensions than to make a portrait from life. Map-makers, after all, were not expected to be animal portraitists, and, in the tradition of the miniature painters and manuscript illustrators, in most cases they took their decorative elements, such as animals and trees, from patterns already extant in emblem- and pattern-books. In some cases, however, map-makers, as well as book illustrators and engravers, had to depict an animal never before represented. Where possible, they would draw from a live specimen or a skin, but in many cases they had to rely on written reports from eyewitnesses or perhaps a scrap of hide, a beak, or a feather. Their attempts to depict the new illustrate both the medieval mentality that Raven suggests underlies the common culture of the Renaissance, and some of the first attempts to use illustration as datum as well as symbol. By examining a number of images showing North American animals, and following their transformations and their use from the sixteenth through the seventeenth century not only in maps, but also in printed accounts of exploration and in the first great compilations of a new group of natural historians, we can begin to understand better the work of these early artist-illustrators.

The problem faced by early artists and cartographers in dealing with new animals is exemplified by the first depictions of the indigenous fauna of the western hemisphere for which there are no Palaearctic or existing exotic equivalents (such as an elephant or a lion, which had been illustrated in manuscript since early medieval times). The works of André Thevet (1502–1590), monk and Cosmographer Royal of France, are the source for a number of first images of the New World flora and fauna, and it is instructive to understand both his sources and his interpretation of them. Unlike many other authors, Thevet had some first-hand experience of the New World, having made a trip to Brazil in 1551, where, however, he spent most of the time ill in bed. On the return voyage he probably coasted North America. He subsequently published two books on the New World, *Les Singularitez de la France Antarctique* in 1558, and *La Cosmographie universelle* in 1575, which are devoted primarily to descriptions of South America, but do include chapters on the northern voyages. He was also the author of a manuscript, 'Grand Insulaire,' which may pre-date the published materials.[47] Thevet has been villified both by his own contemporaries and by later historians for his 'pretentious verbiage, careless composition, clerical errors, and occasional pure fiction,'[48] but Ganong has found nuggets of truth in Thevet's descriptions of 'la terre de Canada,' and he suggests they are the result of Thevet's acquaintance with Jacques Cartier. In *La Cosmographie universelle,* Thevet claims to have stayed five months in the house of Jacques Cartier at St Malo in Brittany, and acted as a journalist to the older explorer, recording his conversations, 'albeit hampered by unfamiliarity with the strange places, objects and events concerned, by the reporter's desire to make a good story for his readers, and by personal limitations.'[49]

Some of the personal limitations that Ganong refers to are in many ways also the limitations of the age. Thevet is treating phenomena that have no point of relation to anything that has gone before. The novel animals and plants of the New World are not in Pliny's *Natural History.* The exotica of the Middle Ages – the elephants, lions, tigers, popinjays, and crocodiles – were also the exotica of the Roman Empire, but the productions of the New World are new. There are things in Thevet's books which, as M. de la Porte says in the advertisement to the reader at the beginning of *Les Singularitez,* appear more monstrous than natural but are yet the products of Mother Nature: 'I have no doubt, Reader, that what is described in this present history will surely make you marvel, as much for the diversity of things here presented to you, as for

PLATE 5: Possibly Jean Cousin the Younger (*ca* 1525–*ca* 1594), Su, 1558. Woodcut. From André Thevet, *Les Singularitez de la France Antarctique, Autrement nommée Amerique: & de plusieurs Terres & Isles decouvertes de nostre temps* (Paris, 1558).

the many others which at first glance will seem to you more monstrous than natural. But having advisedly considered the great purposes of our Mother Nature, I believe firmly that such an opinion will have no more place in your mind.'[50] It is to Thevet, however, that we owe the first published illustrations of the tree sloth, the anteater (or perhaps ground sloth), and the toucan, plus an interesting (though not the first) image of the bison. These illustrations acted as patterns or icons for other artists, cartographers, and engravers. Gesner, in his pandect, referred to Thevet as an authority, and the de Bry family, in their famous series on America, appear to have copied some of his illustrations.

Thevet is the first author to provide an illustration of the 'succarath,' or 'su' (plate 5). In *Les Singularitez,* he describes it as native to Patagonia, but, in *La Cosmographie,* the su has migrated to Florida. (The lack of geographical precision by Europeans in dealing with the northern and southern hemispheres is reminiscent of the nineteenth-century idea of 'darkest Africa,' all equally jungly and fabulous from one coast to the other.)

Thevet decided to include a picture of the su on account of its unusual appearance. He describes it this way: 'There is also found another kind of beast (which for its rarity as well as its deformity I should much dislike to omit) named by this peoples *Succarath* and by the cannibals *Su*. That animal mostly inhabits the banks of rivers and is rapacious, and is very strangely built, such as you see pictured ... If it is pursued it takes its young on its back and covers them with its tail which is rather long and wide, and flees.'[51] Thevet did not draw the su or the other animals himself. He hired an artist, presumably the Mannerist painter Jean Cousin the Younger (*ca* 1525–*ca* 1594), who made the drawings for a number of woodcuts in *Les Singularitez* and those in *La Cosmographie*, and an engraver, both of whom probably worked from a skin, since he mentions the Native people hunted the su and captured it in pit traps. What is most remarkable about the drawing is the face, which looks a little like a monkey's, and the great plumed tail that covers the young riding on its back, suggesting that the su might be an anteater. This strange animal is included in later editions of Konrad Gesner's great encyclopedia of animal life and in Topsell's re-edition of Gesner, where it has gained a reputation for fierceness. Topsell relates that he has inserted 'the true image thereof as it was taken by Thevetus,' and describes it as 'of a very deformed shape, and monstrous presence, a great ravener and untamable wilde Beast.'[52] Unlike Thevet, he calls the su 'cruel, untamable, violent, ravening, and bloudy,' a reputation based perhaps on its cry, which Thevet calls 'terrible,' and on its reported habit of killing and maiming its offspring '(as if maddened)' when caught. Here the words would appear to have served to transform the drawing from that of a peculiar animal to a ferocious one. Whether or not the su deserved its reputation, it appears fierce and altogether more lion-like in Matthaeus Merian's 1630 engraving of a New World landscape, discussed in more detail below. Thevet's su also appeared on a number of maps, including a 1592 chart by Petrus Plancius, where it was located in the far north of Canada.[53] It appeared as late as 1697 in a copper re-engraving in the third edition of a book by Aldrovandi's pupil Gaspar Schott, *Physica Curiosa sive Mirabilia Naturae et Artis*, first published in 1662.

Thevet also provided one of the first printed illustrations of the bison (plate 6) in the chapter on Florida in *Les Singularitez*. Here Thevet likely

PLATE 6: Possibly Jean Cousin the Younger (*ca* 1525–*ca* 1594), Bison, 1558. Woodcut. From André Thevet, *Les Singularitez de la France Antarctique, Autrement nommée Amerique: & de plusieurs Terres & Isles decouvertes de nostre temps* (Paris, 1558).

provided his artist or artists with oral or textual descriptions and some material evidence. Thevet probably had access to the published accounts of the early Spanish explorers who saw the bison in their original habitats in the southern part of North America. Cabeza de Vaca travelled across Florida, Texas, and Arizona between 1528 and 1536. Oviedo y Valdés published Cabeza de Vaca's description of the bison in his *Historia general y natural de las Indias* of 1547: 'They have small horns like the cows of Morocco; the hair is very long and woolly like a rug. Some are tawny, others black ... They come as far as the sea coast of Florida from a northerly direction ranging through a tract of more than four hundred leagues ...'[54] A buffalo also appeared on a 1551 Gutierrez map, and in a book published in 1553 by Francisco Lopez de Gómara to which Thevet may or may not have had access. Thevet claimed as well to have acquired two horns – 'I brought back two, which I still have in my den at Paris [from] when I came back from my first voyage'[55] – and noted in his description of the beast that he had seen a skin:

> ... among others you can see a kind of large bull, having horns only a foot long, and on its back a tumor or hump like a camel's. The hair is long all over the body, whose color is close to that of a tawny mule, and the hair under his chin is even more so. They once brought two live ones to Spain; I saw the skin only of one of them, and they cannot live long there. They say that this animal is the perpetual enemy of the horse and cannot bear to have one near him.[56]

In 'Grand Insulaire,' which was not published, Thevet adds that the 'Buffol beste majestueuse' has a tail 'like that of a lion,' and relates it to the 'bison' of western Europe: 'this species is to be found in Lithuania and Poland which they call Zubex, and the Tartars Roffert.' He also attributes a medicinal property to the horns, and says that 'the barbarians keep them as protection against the poisons and vermin which they often encounter going through the country to do their fishing.'[57] Thevet's 1558 illustration of the bison is somewhat improved in the 1575 *Cosmographie*, which shows Native people hunting bison with spears and with bows and arrows.

Thevet was not the first to depict the buffalo, and it would appear that the earlier Spanish woodcut was the model for at least two other later-sixteenth- and early-seventeenth-century engravings. Gómara's buffalo was similar to Thevet's in that it had a long 'beard' and what appears to be an udder. Later seventeenth- and eighteenth-century engravings are almost universally of males, so that it is likely that Gómara made have had access to an actual female specimen. The historian of the buffalo, F.G. Roe, notes that, among Native people as well as Europeans, there was a unanimous preference for the skins of cows, since bulls' hides 'were practically unworkable.' The meat of the cow was also considered more palatable. What is stunning is the quantity of buffalo and cattle hides shipped to Spain in the sixteenth century. Roe notes that, by 1588, Acosta states that 64,350 hides were shipped to Spain, of both 'tame and wild kine,' so that it is not surprising that Thevet and others saw buffalo hides.[58] De Bry's engravers in 1595 use the Gómara buffalo as their

PLATE 7: Buffalo, 1633. Engraving. From Johannes de Laët (1593–1649), *Novus orbis, seu Descriptiones Indiae Occidentales, libri XVIII* (Leyden, 1633), 56.

model in a map of Mexico, where it is called 'Vacca Indica,' the Indian cow,[59] as do de Laët's engravers in 1630 (plate 7). What is more surprising, perhaps, is that the buffalo does not appear in either Gesner's encyclopedia or Topsell's 1658 *History of Four-footed Beasts ...*, which includes the European bison (or 'wisent') and the 'white Scotian Bison.'

Another animal which also has a long history in print as a monstrous production of the New World is a marsupial animal, the opossum. The transformations of the simivulpa, or apish fox, as it was known in the sixteenth and seventeenth centuries, provide additional insight into the role of the icon-image as used by book illustrators and artists. It is no surprise that the early explorers of South America encountered the opossum. According to mammologist Adrian Forsythe, 'in South and Central America, some sort of opossum can be found in almost any habitat, from wet marshy areas to dry savannah to the edge of the tree line high in the mountains.'[60] North America has only one species of opossum, the Virginia opossum, which extended its range after the last ice age and has only in the mid-nineteenth century, since the arrival of European settlers and the elimination of many of the larger predators, moved northward into Canada. In 1499, Vicente Yánez Pinzón captured a female opossum with young from the trees of Venezuela and brought it back to Spain. The young died on the way, but the adult survived long enough to be described by Pinzón as 'a strange monster, the foremost part resembling a fox, the hinder a monkey, the feet were like a man's, with ears like an owl; under whose belly hung a great Bag, in which it carry'd the young, which they drop not, nor forsake till they can feed themselves.'[61]

The first illustration of an opossum, or apish fox, appeared as a woodcut on a 1516 *Carta Marina* by Martin Waldseemüller. The cartographer might have developed the illustration from the verbal description alone, but the accuracy with which he has drawn the head and ears would suggest that he might have seen the skin or else a drawing of the beast brought back by Pinzón or another. He has not, however, made the hind end to look like a monkey, by which Pinzón probably meant the opossum had a long prehensile tail. (Captain John Smith in 1600 describes the Virginia opossum as having 'a taile like a Rat.')[62] Nor has he made the feet look like hands, even though the opossum's feet do have distinct digits, with an opposable thumb, and thus resemble human hands. The opossum has been described in relation to the familiar, but what both the explorer and the artist had difficulty grasping was the marsupial pouch. It is referred to as a 'great Bag,' and as such it appears on every drawing. What might also suggest that Waldseemüller had at least seen some representation of an opossum is that he includes very prominent nipples. The female opossum does in fact possess elongated nipples, so that the young can suckle and yet lie outside the pouch. The Waldseemüller opossum has a long career as a graphic representation. Konrad Gesner included it, the first New World animal, in the 1551 edition of *Historia Animalium*, which was copied by Topsell in his 1658 *History of Four-footed Beasts* ... Gesner's engraver has added sucking young and reduced the number of the opossum's toes to three. Despite the text noting that 'it hath a tail like a Munkey,' the illustration shows a short-tailed animal. Gesner/Topsell describes the animal in words similar to those of Pinzón:

> ... in the forepart like a Fox, and in the hinder part like an Ape, except that it had mans feet, and ears like a Bat, and underneath the common belly, there was a skin like a bag or scrip, wherein she keepeth, lodgeth, and carryeth her young ones ...: It hath a tail like a Munkey: there was one of them with three young Whelpes taken and brought into a ship, but the Whelps died quickly: the old one living longer was brought to *Sivill*, and

PLATE 8: Matthaeus Merian (1593–1650), New World Landscape, 1630. Engraving. From Johann Theodor de Bry, *Les grands voyages*, part XIV, 55 (Hanau, 1630). Merian's malevolent landscape, with exploding volcanoes, strange beasts, and venomous serpents, has suggested to some that the New World was part of a lesser creation, unfit for civilized habitation. In the background naked Native people wander in what would be a pleasant valley were it not for the erupting volcanoes, spewing fire and brimstone. A strange single-horned beast resembling the camelopardus, but perhaps here a unicorn, stands in the background by a running cat-like animal. In the foreground is a veritable bestiary of strange creatures, including a crocodile or alligator, an armadillo, two lizard-like monsters, two snakes, a scorpion, perhaps two jaguar, what might be an oppossum, and a simivulpa and a su, both likely drawn after Gesner.

> afterward to *Granado*, where the King of *Spain* saw it, which soon after by reason of the change of aire and incertainty of diet, did also pine away and die.[63]

Marcus Gheeraerts (*ca* 1519– before 1604) seems to have based his engraving of a bearish, three-toed simivulpa on the Gesner version, which appears also to have been used as the model in an engraving of a mythical landscape by Matthaeus Merian (1593–1650) in the 1630 edition of de Bry's *Grands Voyages*, pt. XIV (plate 8). These two early-seventeenth-century engravings are of interest for their continuing use of the animal as emblem. Marcus Gheeraerts's simivulpa was one of two almost-heraldic beasts on his allegory of *America*, engraved by Philipp Galle of Antwerp between 1590 and 1600. The other is a goat, which is often featured with the Asian fat-tailed sheep in depictions of the New World. A variety of parrots are included, as well as two snails. Gheeraerts's allegory is interesting in that, unlike a vast number of allegorical works on America, it includes not only Native people in feathered costumes, but also two Inuit, a man and a woman, in what appears to be an interpretation after the original watercolour studies of 1577 by John White now in the British Museum, though Hugh Honour suggests that Gheeraerts might in fact have seen the Inuit himself during a stay in England.[64] Gheeraerts has provided the Inuit woman with a weapon in the form of a club more often associated with the Brazilians, something certainly not in the White drawing or used by the Inuit. Merian's landscape shows what might be termed 'the monstrous productions of the New World' against a background of exploding volcanoes, which are also featured on maps of Central America. Along with the apish fox are other equally bizarre creatures, including a spotted cat (a jaguar?); twisting, venomous snakes of heraldic appearance; scorpions; alligators; armadillos; a one-horned, giraffe-like animal; and what might be an iguana and an anteater. Balancing the simivulpa on the left of the engraving is the baleful su, also with its whelps.

Like the Dutch flower painters of the same period, these artists relied on a set of images that they manipulated to fit the demands of the required pictorial image. Such animals as the su, bison, and simivulpa stood for the exotic, and their transformation into caricatures speaks to the value placed on accurate rendering by sixteenth- and seventeenth-century readers. William Ashworth suggests that, despite their origin as non-emblematic images, their very persistence and reuse render them emblematic. They become, in fact, emblems of first-hand observation.[65] This respect for the original representation as embodying an eye-witness account is so strong that, on occasion, as we have seen with the simivulpa, it overwhelmed the textual description, which also purports to be that of an eye-witness. Text and image companion each other, but they cannot be said to act as glosses for each other. The intimate relationship between word and picture awaited another generation of illustrators, in the eighteenth century.

two

NATURALISM AND THE COUNTERFEIT OF NATURE

We have examined the use of animals as signs or marks for the new lands being explored and the development of new iconic images by authors and artists struggling to represent a creature for which they had no model. We have also taken a cursory look at some of the transformations of the iconic image of a particular creature over a period of a century or more. We are left with the impression that the visual world of the sixteenth and early seventeenth centuries, at least as far as the depiction of the natural history of the New World was concerned, was dominated by either simple cartoon drawings or misshapen fantasies. This was not, of course, the case. The Renaissance marked a transformation from the medieval world to the modern, and the changing nature and use of images reflected this transformation. The increasing importance of naturalism and the relationship between what is visible and what is known are interwoven in this period. Looking at what to our eyes appear relatively crude manuscript or map illustrations and even cruder woodcuts does not prepare the contemporary viewer for the delicacy and realism of fifteenth- and sixteenth-century painting, particularly watercolours. The accuracy that seemed not to be overly prized in book or map illustration was certainly possible and often desired. The author Fernandez de Oviedo y Valdés, for example, longed for the skills of famous painters to render the wonders of the New World: 'It needs to be painted by the hand of a Berruguete or some other excellent painter like him, or by Leonardo da Vinci or Andrea Mantegna, famous painters whom I knew in Italy.'[1]

The present-day wide availability in printed form of paintings and drawings from the Renaissance should not obscure the fact that, for most people living in the fifteenth and sixteenth centuries, the only access to images was through paintings on display in churches and palaces, or reproduced as woodcuts or engravings in printed books. Even if Leonardo or Mantegna had been available to paint the productions of the New World, it is unlikely that the readers of Oviedo y Valdés's 1535 book would have been able to appreciate his skill. An examination of some of the original works of art that depict the natural history of the New World, and of their transformation into printed information, can illuminate the space between artist and reader.

The Development of Naturalism

Animals appear on manuscript maps as symbols of place, and, as David Knight points out, they also have a place in 'symbolic science.' The animals in bestiaries or later emblem-books did not require accurate depiction: 'Once their essence was known, there was little need to study them further or to draw them more accurately than the accepted image, which after all called up the required associations.'[2] Once the graphic characteristics of the representation had been determined, the animal could be copied again and again without recourse to the original, which often in the case of lions, eagles, elephants, or even salamanders, might be hard to come by.[3] Lack or scarcity of living models, however, was not necessarily the reason for the lack of realism found in the depictions of animals, either exotic or familiar. Francis Klingender suggests that 'to understand how utterly remote these draughtsmen were from nature one need only turn to their illustrations of creatures they might easily have observed in the monastery garden ...,' such as the hedgehog. Certainly the hedgehog featured in Villard de Honnecourt's thirteenth-century sketchbook is a mere bristly caricature of the common animal. Klingender points out that 'the crudest representation of a creature might be sufficient so long as it drew attention to those selected characteristics which were already associated in the reader's mind with accepted mystical, symbolic or moral ideas. A convention was thus established in which animal pictures become a kind of pictorial shorthand ...'[4] The same applied to early printed herbals based on classical manuscript models, whose increasingly debased illustrations followed, not the living plant, but the text.[5] The fact that medieval illustrations of plants and animals often appeared crude and inaccurate did not mean that the artists necessarily saw this way. Any acquaintance with manuscript illumination or fourteenth-century artists' sketchbooks would dispel this view.

Paintings were often 'built up' by artists from drawings in sketchbooks or pattern-books. A sketchbook now in a Cambridge University collection was used as a pattern-book in an English artist's workshop at the end of the fourteenth century, and includes labelled sketches of well-rendered birds, often, however, copied from even earlier medieval sources.[6] Bird and animal designs were much in demand not only by artists and illuminators, but also by embroiderers.[7] Italian sketchbooks of the same period also feature naturalistically rendered watercolours of animals observed from life. The idea of painting from nature, however, was still novel, and the fourteenth-century artist Giotto was considered a child prodigy, in that he was discovered sketching a sheep from life while he was tending the flocks. Drawing from life was not considered an end in itself. Otto Pächt notes that the drawings in the sketchbooks were 'still or stilled life rather than life.' Although an artist like Giovanni de'Grassi does break with the 'medieval habit of abstract invention ... [and] now finds his models in nature itself, yet [he] at once turns the newly discovered material into patterns, similes, atelier formulas.'[8]

This subordination of accurate rendering to pattern has convinced Klingender that there is little relationship between these pattern-book and sketchbook studies and 'the naturalism of the fifteenth century, which was linked with the renewed scientific revival of the Renaissance.' It is the development of a new vision, based on the mastery of perspective, that allowed fifteenth-century artists to lift the animal off the page and out of the pattern to become a portrait of a living thing. By the sixteenth century the 'scientifically controlled projection of a three-dimensional world ... came to be accepted in Europe ... as the only mature mode of vision.'[9] Naturalism as a mode, however, was in the early period still a matter of details, of beautifully rendered flowers, or minutely portrayed insects, often set against a blank background. The conventions of naturalistic portrayal could, however, answer to the needs of writers like Oviedo y Valdés, who longed for the accurate, 'true-to-life' rendering of the wonders that they saw in the New World and of the objects that they brought back to Europe. That they were not necessarily able to realize this desire was the product of both convention and technology. By briefly examining the work of three sixteenth-century artists, Albrecht Dürer, Jacopo (or Giacomo) Ligozzi, and John White, the sources for their life studies, and the transformation that their original images undergo in the translation to print, we will come closer to understanding the opportunities and constraints that affected not only the artistic, but also the scientific community in the sixteenth and early seventeenth centuries.

Albrecht Dürer (1471–1528) is perhaps the best known of the German school and a master of drawing, painting, woodcut, and engraving. Dürer's woodcuts and engravings are most familiar, but it is instructive as well to look at his watercolour studies of plants and animals. Dürer is justly famous for his highly detailed renderings of flowers and his studies of animals. He advised artists that 'life in Nature showeth forth the truth of these things, wherefore regard it well, order thyself thereby and depart not from Nature in thine opinions, neither imagine thyself to invent aught better ... For Art standeth firmly fixed in Nature, and whoso can rend her forth thence, he only possesseth her.'[10] In many ways watercolour drawing is ideally suited to rendering Nature. Svetlana Alpers notes that watercolour 'was primarily employed in the interest of immediacy of rendering.' It is a medium 'that allows drawings to display at once two normally contradictory aspects: drawing as inscription (the recording on a surface) and drawing as a picture (the evocation of something seen).'[11] In the case of the columbine, the great piece of turf (*Das Grosse Rasenstück*), the famous rabbit, or the study of the Blue Roller,[12] the use of watercolour allows Dürer to depict in meticulous detail the exact appearance of the particular thing. These are certainly paintings from life, but they are not yet paintings of life. The plant or animal appears isolated on a ground, like a drawing in a sketchbook, and indeed Dürer did work some of these watercolour studies into other works. His iris, for example, appears in his painting *The Madonna with the Iris*. More significant, perhaps, is Dürer's understanding of the way in which 'a well-practised artist' might work. Such an artist 'hath no need to copy each particular figure from life. For it sufficeth him to pour forth that which he hath for a long time gathered unto him from without ...'[13]

The approach to the object which Dürer brought to his images of natural history differed little from that of the two later sixteenth-century artists. Dürer did most of his watercolour studies in a brief period at the beginning of the sixteenth century. Both Jacopo Ligozzi (1547–1626) and John White (fl. 1577–1593) worked in the 1580s. Ligozzi never travelled to the New World, but he was esteemed as a painter of animals and plants. He became Court

Painter to the Duke of Tuscany and was later Superintendent of the Uffizi Gallery, but his most rewarding collaboration as far as natural history was concerned was with Ulisse Aldrovandi (1522–1605), founder of a natural-history museum and herbarium and botanic garden in Bologna, and author of a number of natural-history encyclopedias, from the *Ornithologiae* (1599) to the *Dendrologia* (published in 1668, long after his death). Ligozzi supplied many of the original illustrations for these books, though the reproduction of his precise paintings left something to be desired. Ligozzi prepared a number of watercolours of New World species, including the first lifelike picture of an agouti and a splendid parrot, which may have come from Aldrovandi's museum or the Duke's vivarium. Among his botanical studies were a pineapple, an agave, and the 'Marvel of Peru,'[14] all of which were recently cultivated as exotics in Italy.

While Ligozzi worked from specimens imported to Europe, a few artists, like John White, recorded specimens in the field. White prepared his watercolours directly from nature during his tenure as official artist to the 1585 colonizing expedition to Virginia led by Sir Walter Raleigh. The tradition of having an artist in the field did not, however, begin with the American voyages of discovery. The author of the *Gart der Gesundheit*, a German herbal printed in 1485, realizing 'that many noble herbs did not grow in this German land,' went on a spiritual and botanical pilgrimage to the Holy Sepulchre, taking with him 'a painter of understanding and with a subtle and practised hand' to paint and draw the herbs in 'their true colours and form.'[15] The next year, the author of a printed travel book to the Holy Land, Bernard von Brudenbach, Bishop of Mainz, took with him 'an ingenious and learned painter,' Erhard Reuwich, to record the scenery and animals 'truly depicted as we saw them in the Holy Land.'[16] White was not even the first official artist to travel to the New World. Jacques le Moyne de Morgues (d. 1588) had been the artist of a colonizing expedition to Florida by René de Laudonnière in 1563–5, and the physician Francisco Hernández (1517–1587) spent six years (1571–7) assembling notes and drawings on aspects of Mexican natural history for Philip II of Spain. The Spanish example stimulated Richard Hakluyt (1552–1616) to insist that 'a skilful painter is also to be carried with you, which the Spaniards commonly used in all their discoveries to bring the descriptions of all beasts, birds, fishes, trees, townes etc.' Apparently Sir Francis Drake, on his circumnavigation (1577–80), heeded this advice and 'kept a book in which he entered his navigation and in which he delineated birds, trees and sea lions. He is adept in painting and has with him a boy, a relative of his, who is a great painter.'[17] White knew le Moyne de Morgues, who had settled in London, and was probably acquainted with Baptista Boazio, the artist for Drake's voyage in 1585–6. While White's instructions as official expedition artist do not survive, it is likely that they were similar to those prepared for Thomas Bavin, who accompanied Sir Humphrey Gilbert to New England in 1582. Bavin was instructed that he was to 'drawe to lief one of each kinde of thing that is strange to us in England ... all strange birdes beastes fishes plantes hearbes Trees and fruictes ... also the figures & shapes of men and woemen in their apparell as also their manner of wepons in every place as you shall finde them differing.'[18] The artist and his observer were to be attended by others to carry their equipment, and they were also to make 'cardes' or maps depicting with symbols the variations in landscape, vegetation, location of villages, and so on. For the 1585 expedition, White worked with Thomas Harriot, who recorded all observations, later published by Theodor de Bry as *A briefe and true report of the new found land of Virginia* (1590) with engravings after White's original drawings.

White had been trained as a limner in England, and was familiar with the miniaturist tradition, and the importance of making a likeness of the person or object to be painted. He also had likely some experience in travel to the New World as a member of Frobisher's second expedition to the Arctic in 1577. White was certainly the artist who depicted three Baffin Island Inuit brought back to England by Frobisher, and his watercolour renderings were much copied in books and engravings (see the previous chapter for a description of the engraving by Philipp Galle). Whereas Oviedo y Valdés had wished for an artist with the skills of a Leonardo, Raleigh at least had someone with sufficient training to depict with a high degree of naturalism the productions of the New World. Only sixty-five of White's original watercolours have survived. Some were destroyed as they departed from Roanoke Island, when 'most of all wee had, with all our Cardes, Bookes and writings, were by the Saylers cast over boord' in bad weather.[19] Others were lost in the sixteenth or seventeenth century, but a set of copies was discovered by Sir Hans Sloane in the early eighteenth century and thus came into the collections of the British Museum. Among these copies and the originals, there are watercolour sketches of a variety of fish, birds, some insects, several lizards and invertebrates, three wonderful turtles, and a very few plants. Harriot notes that, even had they wanted to, they could not manage to depict all the plants and animals. He writes 'we haue taken, eaten, & haue the pictures as they were there drawne with the names of the inhabitants of severall strange sorts of water foule eight, and seventeene kinds more of land foul, although we haue seene and eaten of many more, which for want of leasure there for the purpose could not be pictured ...'[20]

The Representation as Counterfeit

Original watercolours of plants and animals painted in the conventions of the new naturalism were used in a manner quite different from that of the artists who had relied on pattern-books to add an animal to a painting or tapestry design. The illustration or image of the animal became, not a cypher in a pictorial shorthand, but a 'true' representation of the thing itself. By the end of the sixteenth century, an Italian author could write in *Tractate Containing the Artes of curious Paintinge* (1598) that 'painting is an arte; because it imitateth naturall thinges most precisely, and is the *Counterfeiter* and (as it were) the very *Ape* of Nature: whose quantity, eminencie and colours, it ever striveth to imitate ... by the helpe of *Geometry*, *Arithmeticke*, *Perspective*, and *Naturall Philosophie*, with most infallible demonstrations.'[21] The publication of Otto von Brunfels's (1464–1534) herbal *Herbarum Vivae Eicones* in 1530 with illustrations by Hans Weiditz (d. *ca* 1536), Dürer's pupil, emphasized even in its title the idea of 'living images.' (The herbal tradition is examined in chapter 3.) The image also became a medium for the transmission of information; it stood for the specimen. Given the present-day ubiquity of printed images, we take for granted easy access to the visual. In the sixteenth century, acquiring an image necessitated the use of an artist, a painter or limner, whose skill might vary from that of a Ligozzi or White to that of a hack. Ulisse Aldrovandi, the sixteenth-century pandect author, worked with a number of artists besides Ligozzi and ran even his local collecting trips like expeditions, taking both draughtsmen and secretaries with him when collecting outside the city, thus guaranteeing the accuracy of both the image and the notes. Whatever the quality of the original painted image, it became for the naturalists of the sixteenth century a medium of communication.

According to Lynn Thorndike, the sixteenth century saw the rise of a novel phenomenon in the world of learning – 'cooperation between different individuals. They send one another specimens or at least drawings or written descriptions of strange animals and unfamiliar herbs which they have run across ... This cooperation in science by men of diverse nationalities, professions, religions and even philosophies is indeed impressive.'[22] The increased supply of printed books had changed the nature of scholarship, encouraging isolated scholars to compare notes on the printed texts in a manner that had been impossible in the age of manuscripts. At the same time this new practice of correspondence favoured not only the exchange of textual criticism and written commentary, but the exchange of images. The need for accurate representations and the flood of hand-painted images (chiefly watercolours) that this correspondence produced has not been sufficiently appreciated. A scholar like Konrad Gesner benefited a great deal from this flow of correspondence and the dried specimens or painted images that often accompanied it. Gesner acknowledged the importance of illustrations 'so that students may more easily recognize objects that cannot very clearly be described in words,'[23] and his correspondents fed his appetite for paintings and specimens. Gesner remarks in *De Herbis Lunariis* (1555) that 'fifteen years ago or thereabouts ... an Englishman returning from Italy greeted me ...; among other pictures of the rarer plants which he had taken care to get painted, he showed me a picture of a Elleborine ...'[24] The Englishman was the naturalist William Turner (1508–1568), to whom, Gesner notes in *Historia Animalium*, he in return sent pictures. Gesner was also in correspondence with John Cay, or Caius (1510–1573), author of *De Rariorum Animalium atque Stirpium Historia* (1570), who provided him with a number of portraits of animals in their proper colours, including those of a lynx and a civet cat lodged at the Tower. The entry under the 'ZIBETH, or SIVET-CAT' reads: The best description of this beast in all the World, that I could ever finde, was taken by Doctor *Cay*, and thus sent as it is here figured to Doctor *Gesner* with these words following. There came to my sight (saith Doctor Cay) a *Zibeth* or *Sivet* very lately, which was brought out of *Africa*, the picture and shape whereof in every point I caused to be taken, which is this prefixed, so that one Egge is not more like another, then this is to the said Sivet or Zibeth.[25] Cay also corresponded with Gesner concerning the marmoset, which is 'much set by among women.' Gesner's *Historia Animalium* included an illustration of the marmoset ('This figure of the *Sagoin*, I received of *Peter Cordenberg*, a very learned Apothecary at *Antwerp*'), but '*John Cay* that famous *English* Doctor hath advertised me, that it no way resembleth the *Sagoin* it self ...'[26] Towards the end of his life, Gesner received a shipment of dried plants from Jean Bauhin (1541–1613), brother of Gaspard (1560–1624), which he spent the summer having 'painted from life' so that he could return them to their owner.[27] There still exists a number of Gesner's own annotated watercolours, including one of a tobacco plant painted *ca* 1554–5. The plant is carefully drawn and coloured from roots to blossom, with sketched details of flowers, leaves, and fruit.[28] Thomas Penny (b. 1530), the English entomologist, also collected information on insects from his correspondents on the Continent, receiving drawings from Johann Camerarius the Younger (1534–1598), author of a 1590 emblem-book, and a picture of a Goliath beetle from the Duke of Saxony's museum.

John White's unique watercolours were in obvious demand by naturalists and compilers both during his lifetime and after. Penny received watercolour sketches of four New World insects, including a beautifully executed tiger swallowtail, directly from John White. White evidently also knew John Gerard (1545–1612) who made an engraving of White's watercolour of a milkweed (*Wisakon*) for his *Herball*. White's illustrations were also copied into the unpublished manuscript version of Edward Topsell's *The Fowles of Heaven* (1608–14), an incomplete translation of Aldrovandi's *Ornithologiae*. Topsell (1572–1625) had received about ten copies of bird sketches by White, perhaps from Hakluyt, and these constitute his only significant addition to Aldrovandi's work.[29] A century and a half later, Sir Hans Sloane's copies of White's originals were consulted by Mark Catesby, who copied no fewer than seven items directly into *The Natural History of Carolina* ... (1729–43).

These are but a few examples of the widespread exchange of pictorial information among naturalists. Prior to the development of well-preserved museum and herbarium collections that could be visited and studied, naturalists often depended on the naturalistic image passed about among a circle of acquaintances.[30] Collections of illustrations became in effect a portable museum, and the accuracy of an image taken from life was trusted. In the classification of birds, for example, Anker notes that 'right down to the close of the eighteenth century native and foreign birds were in many instances described and named from unpublished pictures ...'[31] In some cases, the pictures were the only specimens. Wilfrid Blunt notes, for example, that, when ants damaged Nikolaus Joseph Jacquin's herbarium material in the late eighteenth century, 'he therefore supplemented his description and notes on the new species with watercolour drawings. These accordingly are the equivalent of type specimens.'[32] The importance of the original image and its relationship to classification remains undiminished even into the eighteenth century, when the role of vision and of the visible assumed great significance in Linnaean classification.

Translating the Image: The Value of Repeated Pictorial Statement

While sixteenth-century naturalists had learned to trust the manuscript images which they exchanged and classified, the translation of the original image into a reproducible format, which meant either the woodblock print or, by the end of the century, a copperplate engraving, was not without problems. It is these problems of translation of the original image with a high-enough degree of resolution to communicate the intent of the artist that led William Ivins to develop his theories on the print as a medium of visual communication. 'This means,' he writes, 'that, far from being merely minor works of art, prints are among the most important and powerful tools of modern life and thought.'[33] While the adoption of perspective drawing allowed the artists to develop a new means of approaching verisimilitude in their pictorial images, the print, 'the exactly repeatable pictorial statement,' allowed the readers to rely on picturing as well as writing. While it might be argued that the reliance on picturing was the result of the increasing verisimilitude of the new aesthetic, Elizabeth Eisenstein points out that there is a qualitative difference in the accuracy of the sketchbook rendering – the manuscript – and the widespread dissemination of an accurate printed image. In her discussion on the impact of printing technology, she takes issue with the art historians who would suggest that Leonardo da Vinci's understanding of anatomy laid the foundation for the new anatomical science: 'In my view it is an exaggeration to launch modern science with the advent of perspective renderings and to regard pictorial statements as sufficient in themselves.'[34] She suggests that the ease and accuracy of modern mechanical and electronic reproduction techniques have blinded us to the fundamental change in vision and meaning that occurred when pictorial images became printed images:

> we are not accustomed to distinguish between the much-copied hand-drawn

> image in the much-used reference work and the freshly drawn image in the unique sketchbook or manuscript. The visual contrast between the 'fine' pen drawing and 'crude' woodcut is so powerful, the difference between fresh and copied handwork is especially likely to be overlooked. This is unfortunate. The difference between the hand-copied image that decays over the course of time and the repeatable engraving that can be corrected and improved is essential for understanding how visual aids were affected by print ... modern reproductive techniques have been unhelpful and block our vision of an earlier process of change.[35]

While Eisenstein is convincing in her arguments about the new use and value of printed documents, what is not clear in looking at the transformation from manuscript to printed image is the importance of the degree of resolution, that is, how well the printed image reflects the real thing. Naturalists had decided to accept the new naturalistic pictures as simulacra of the real world. Watercolours such as those of Dürer, White, and Ligozzi incorporated certain conventions in their rendering, particularly with regard to perspective and colour, but these, for the most part, disappeared in the acceptance of the new visual language. The degree of resolution of the printed image, however, could be affected by three factors: the technology of printing and colouring; the conventions of the engravers and cutters; and the relationship of the printed image to the text.

Woodblocks were admirably suited to the first generation of printed books. Because they were relief cuts, they could be bound into the press and printed with the type. It was rare, however, that the original drawing and the woodcut were done by the same person. In general, the artist would make the drawings with a quill or a brush on the block. The *Formschneider*, as the cutter of forms was called in German, would then cut away the surface with knife and chisel, producing the image on the block. Theoretically, then, the print was not a translation, but relayed the actual intention of the artist. In practice this was not always the case. Jost Amman, a *Formschneider*, advertised his skills in 1568: 'I am a good woodcutter and I cut so well with my knife every line on my blocks, that when they are printed on a sheet of white paper you see clearly the very lines that the artist has traced, his drawing whether it be coarse or fine reproduced exactly line for line.'[36]

Line-for-line reproduction was possible when lines were few and widely spaced. As the artists demanded greater and greater resolution of their pen-and-ink drawings, the number of fine lines increased, in a process that Ivins calls 'informational pressure.' Finely detailed blocks were very difficult to print, since the block was inked, not by rollers, but by leather balls soaked in ink. With very fine relief lines crammed closely together, splotches were inevitable. In addition, paper was made 'good one side,' so that the impression on the smooth side was more legible than that on the rough. The production of high-quality woodblock prints was thus time-consuming and expensive. Many of the naturalists, including Gesner, Aldrovandi, and even Fuchs, avoided highly detailed cuts, with a subsequent loss of information.

Because producing woodblock prints was labour-intensive, printers relied on shortcuts, often using old blocks or amending them to suit the text. As noted above in looking at the images used in the Columbus letters, a *Formschneider* might simply recut a block to add an image or a name. Gesner's *Formschneider* recut the illustration of the salamander that appeared in the 1486 *Peregrinationes*, though Gesner did add a note to the effect that the image was not a good one ('*Salamandrae figura falsa*'). Topsell's printers reproduced the images from Gesner with a degree of accuracy which suggests that they either used the old blocks or pasted the prints onto blocks and cut through them. In some cases a printer, loath to lose the investment in cutting of a complex woodcut, might simply remove a portion that required modification, inserting a new carved plug into the block, or even simply printing in the blank section. Such would appear to be the case in the 1611 reprint of *The Noble Art of Venerie or Hunting*, in which the figure of King James I is superimposed on plates from the original 1575 edition which had shown Elizabeth I. The imperfect matching of the blocks shows distinctly as white lines in the print.[37] Buying a set of blocks or plates from a printer was practised a great deal, as we shall see when we look at the herbal tradition in the sixteenth and seventeenth centuries. The engraver and compiler Theodor de Bry (1528–1598) obviously frowned on this practice, since in the note to the reader in his edition of Thomas Harriot's *A briefe and true report of the new found land of Virginia* ... (1590), he wrote: 'Finallye I hartlye Request thee, that yf any seeke to Contrefaict thes my bookx, (for in these days many are so malicious that they seeke to gayne by other men labours) thow wouldest giue noe credit unto suche conterfaited Drawghte. For dyuers secret marks lye hiddin in my pictures, which wil breede Confusion unless they bee well observed.'[38] Nevertheless, de Bry and his sons themselves reused their original images, transplanting, for example, the same scene of deer-hunting in the New World from Newfoundland in volume 13 (1628) to Mexico in volume 14 (1630).

Colour is not trivial in natural history. A white bear is very different from a black bear, as Desceliers indicated on his highly coloured maps. Colour in flower or feather is often diagnostic, and we see the pattern of colour in the red-winged blackbird or blue jay long before we define the shape. The naturalists who exchanged watercolour renderings of plants, animals, and fish saw them, for the most part, in high colour. It is a simple-enough experiment to compare the colour photograph of a work of art with its black-and-white equivalent. Even in the works of John White, which have suffered considerably from poor handling and water damage, the colour is luminous, and, in particular, in the silvery moonfish, the swallowtail, and the flamingo, the use of colour is exceptional. Early prints were black and white. They imitated rendering in pen and ink, and shading was not the result of deepening a hue, but of a series of closely spaced lines, or even cross-hatching. This use of line for modelling became even more pronounced in engraving, and Ivins has written extensively on the 'web of rationality' used by the engravers to translate the image.

As artists became more adept at engraving and etching, the lack of colour was not necessarily a shortcoming in artistic depictions, but in the early attempts at naturalistic rendering of plants and animals, colour was wanted and obtained. In general, colour was added by hand. In 1493 Hartmann Schedel published the *Nuremberg Chronicle*, the 'first great picture-book for the bourgeoisie – your money's worth of pictures – sold unbound and uncoloured at two Rhenish florin, bound and coloured at six.'[39] The 'money's worth' was more than 1,800 pictures, and colour was usually laid on with a very broad brush. Some leaves in my collection from a copy of Aldrovandi's *Ornitholgiae* show a lavish and garish use of colour. The *Certhia,* in English a creeper (*Anglis Crepera à reptando teste Turnero*), boasts red feet, a rufous body, and a purple wing. It perches on a branch of a generic tree with bright green leaves and a single red berry. Turner actually describes the creeper as 'having a whitish breast, the other parts dull brown, but varied with black spots.'[40] The *Spipola altera cum Iunco leui* is yellow, with a very bright blue-green wing, orange throat and beak, and orange legs. The *Spipola alba* is all-over reddish with a bright blue wing, which belies its name. The *Iynx mas.* is cerulean blue, with a red tail, legs, beak, and eye. The female, despite being figured separately, is identically coloured.[41] It is hard to say whether these leaves were painted by the publisher or by the owner. Some books came complete with instructions as to colours intended by the author. In an explanatory text for the tenth woodcut of his book, one author suggests that 'if the cut be coloured the Cow may be painted red, since the animal he has in mind is the red Heifer of Numbers XIX ...'[42] David Bland opines that the reason the colouring is often so crude in early woodcuts is that, 'while a skilled artist would take enormous pains over a single initial, only a hack would undertake the task of colouring up the large numbers required in a printed edition.'[43] Not all coloured editions were so crudely daubed, however, as Barker discusses in his splendid volume on the *Hortus Eystettensis.* In the winter of 1613–14, Philipp Hainhofer wrote to his employer, Duke August, that he might obtain the *Hortus Eystettensis* for 35 florins uncoloured and 500 florins illuminated, but that the price was well worth it, as the cost of colouring each plate was approximately 1½ florins.[44] Similarly, a coloured and gilt copy of the de Bry edition of le Moyne de Morgues' account of the French colony in Florida exists in the New York Public Library. The de Bry series of thirteen volumes on America was not intended as a cheap chapbook for the masses. It was read by the European aristocracy and members of the court, by educated people, by collectors, and by the merchants and artisans with an interest in travel and exploration.[45] The de Bry engravers produced fine and highly detailed engraving, which was complemented by a restrained use of colour wash, likely added at the time of purchase. The colour remains, however, very vibrant, the green almost chrome, the red very vivid. Skies are not just blue, they glow with sunset colours. The Native people are a uniform pale tan, with surprisingly bright belts of blue and red, and in one plate (XXIV) use a blue and yellow fan. Animals tend to be all one colour, either green, brown, or grey, and sometimes, in the case of turkeys and fish, with a splash of red.

The conventions of the engravers and printers also affected the way in which the image reflected the original object. Conventions were of two sorts, the technical conventions of the craft, and artistic conventions. Whereas in a watercolour the lines are soft and indistinct, colours shading into one another, in a print, by virtue of the medium itself, they must be hard-edged. Making a woodcut involved cutting away the wood to create raised lines and shapes. Engraving meant making lines in a metal plate. In each case the *Formschneider*, or engraver, added what Ivins called a 'web of rationality' to the original pictorial image. Ivins would also argue that the woodcut, since the artist drew directly on the wood, preserved the artist's intentions, while the engraving, since it was cut into metal by a craftsman following an existing image, was a translation. In both cases, however, the use of line, as opposed to colour washes and varying shading, contributed to distancing the pictorial representation 'from life.' In some cases, the distance was not far; in other cases, it was so great that the value of the printed image as a conveyor of information was negligible. Looking at two examples of the translation of information from one medium to another, one in a woodcut, and one in an engraving, supports this assertion.

We have already discussed the use of 'manuscript' images as a medium of communication among scientists in the sixteenth century. In 1515, Valentin Ferdinand, or Fernandes, a Moravian painter, sent a description and sketch (presumably coloured, but perhaps not; it does not appear to exist any longer) of a new and curious animal from Asia to a friend of Albrecht Dürer's. Dürer was shown the sketch and description, and later recorded an excerpt from Ferdinand's letter on his own original drawing: 'On 8 May, in the year 1513, a big animal, which they call a rhinocerate, was brought from India to Lisbon for the King of Portugal, and as it is such a curiosity, I must send you its likeness ...'[46] (The year was actually 1515; Dürer's error was corrected on a Dutch version of the woodcut.) Dürer never saw the animal in the flesh. It had been presented to Dom Manuel of Portugal by Sultan Muzafan II, called the King of Cambaia, or Gujarat, but it drowned in a shipwreck *en route* to Pope Leo X in Rome. Dürer was obviously intrigued enough by the likeness and the description that he produced his own pen-and-ink drawing of the 'Rhinoceron.' He included some of the textual description on the drawing, likely gleaned from the original letter:

> It is represented here in its complete form. It has the colour of a speckled turtle. And it is almost entirely covered by a thick shell. And in size it is like the elephant but lower on its legs, and almost invulnerable. It has a sharp horn on its nose which it starts to sharpen whenever it nears stones. The stupid animal is the mortal enemy of the elephant ... Because the animal is

> so well armed, the elephant cannot do anything to it. They say the rhinoceros is fast, lively and clever.[47]

Dürer also prepared a woodcut from the drawing. Despite never having seen the animal, Dürer managed in both drawing and woodcut to capture a likeness of the beast so vivid that it became the 'reference' copy for naturalists such as Gesner and Aldrovandi, and other authors and printers up to the mid-eighteenth century. (In 1739, a version of the Dürer rhinoceros was cut in London for a 'Print publish'd immediately upon the arrival of the Rhinoceros in 1739, by Overton without Newgate'; another version attacking an elephant was engraved by Francis Barlow as 'a true representation of the two great masterpieces of Nature, the Elephant and the Rhinoceros, drawne after life, lately brought over from the East Indies to London ...')[48] While Dürer had great skill in making a drawing for the woodcut, Ivins suggests that in preparing the cut 'even so great an artist as Dürer' could not avoid the constraints of the web and the taint of virtuosity that its use implied. Use of the web was similar to the use of the 'net of rationality' by geometers, 'a geometrical construction that catches all the so-called rational points and lines in space but completely misses the infinitely more numerous and interesting irrational points and lines in space.'[49] What Dürer prepared in his drawing was an outline sketch with a series of circles and overlapping arcs to indicate what are in actuality granulations imitating scales. The deep folds in the skin became distinct lines, resembling plates, and the 'well-armoured' idiom was carried through in a fanciful kind of ribbing on the sides and in the flanges and twisted hornlet on the neck. In the pen-and-ink drawing, these lines, circles, and ribs are tentative and lightly shaded, suggestive of skin. In the cut and the innumerable engravings that reproduced it, the markings are definitive; the ribbing hard-edged, forming a boss; the legs definitively scaled. The formalization of the armoured rhinoceros could be carried to a decorative extreme by the printers, as it was in the 1708 *Voyages et avantures* ... of François Leguat. Here are pictured five rhinoceros, based on the Dürer reference model, one of which is covered with ovals and circles representing the 'armour,' but looking more like a painted toy. At this point the web has produced an image closer to that of the emblem-book than of the image 'after life.'

The conventions of the engravers and printers were also affected by the aesthetic mode of the period. The influence of contemporary artistic fashion has been pointed out by many scholars, but is particularly relevant when artists come to portray the unfamiliar. It is readily apparent in the works of John White, for example, in his drawings of an Inuit man and woman made in 1577. Against all reason and in defiance of the visual reality, White indicated the navel on the portraits of an Inuit man and woman, even though it would not have been visible under heavy sealskin clothing. This Mannerist convention is derived from Roman work where the body contours are discernible under light draperies, and this anomaly was repeated in subsequent engravings after the White drawings. It has often been noted as well that early engravers gave to the aboriginal peoples of America a classical look, in keeping with both Renaissance and Mannerist ideas about the nude. Hugh Honour has suggested that the style of the drawing, by the Mannerist painter Jean Cousin the Younger, and the woodcuts in Thevet's *Les Singularitez* may have inspired the poetical descriptions of Brazilians as the inhabitants of a new 'Golden Age.'[50] Certainly when Theodor de Bry prepared the copper engravings from White's watercolours, the classicizing influence was manifest. White himself never depicted an unattractive Native person, but de Bry's engravings sharpened the features (which in White's drawings were often portraits) into classical heads with a European cast. The bodies as well became more muscled and well proportioned, the hands and feet more delicate, and the hair attractively knotted. De Bry did not engrave White's animal and plant drawings per se, but he did incorporate some of them into composite engravings showing landscape, and it is worth examining both these plates and the manner in which de Bry went about his work in creating the first mass-produced picture book about the New World.

Most Europeans became acquainted with White's vision of Virginian Native people and wildlife through the engravings prepared by the Flemish publisher Theodor de Bry for his 1590 publication in four languages of Thomas Harriot's *A briefe and true report of the new found land of Virginia*. This was the first volume in de Bry's series *America*, and was followed by a publication based on the le Moyne drawings of Florida. De Bry arrived in England in 1587 to prepare some engravings. Here he became acquainted with both the work of White and le Moyne and acquired sets of their original drawings. De Bry described his enterprise in the epistle 'To the gentle Reader' in the English version of Harriot:

> Consideringe, Therfore that yt was a thinge worthie of admiration, I was veryе willinge to offer vnto you the true Pictures of those people wich by the helfe of Maister Richard Hakluyt of Oxford Minister of Gods Word, who first Incouraged me to publish the Worke, I creaued out of the verye original of Maister Ihon White an Englisch paynter who was sent into the contreye by the queenes Maiestye, onlye to draw the description of the place ... I craeued both of them [the White and le Moyne drawings] at London, an brought Them hither to Franckfurt, wher I and my sonnes haven taken ernest paynes in grauinge the pictures ther of in Copper, seeing yt is a matter of noe small importance.[51]

De Bry did take pains to remain largely faithful to White's originals, though the transformation effected in the translation to print went far beyond a recasting into a Mannerist aesthetic, and hints at another convention of the artists and engravers that made the representation of the actual more problematic. Plate XIII, *Their manner of fishynge in Virginia*, is one of the few in the Harriot book to show wildlife in a more than incidental way (plate 9). It features a number of fish and aquatic invertebrates as well as birds, and the comparison with the original watercolour is instructive.

In the original watercolour, White has shown a dugout *cannow* with four Native people, two standing and two sitting by a fire in the middle of the craft. In the background is shown a fish weir and two other Native people spearing fish. In the far distance is another canoe, and in the sky a number of birds. Hulton notes that this is a composite drawing combining night fishing and day-fishing techniques and a number of species, including an anomalous West Indian hermit crab. The fish shown, however, are recognizably a catfish, burrfish, hammerhead shark, skate or ray, and perhaps a sturgeon. There is also a depiction of a rather peculiar and poorly drawn horseshoe crab (often erroneously called a 'king crab'). A pelican, two swans, and a flight of ducks are sketched in the sky. The engraving appears as the double-page plate XIII in the 1590 German edition of Harriot.[52] All the plates in this volume are gathered at the back and the text is printed along the side. In the engraving, the fish have multiplied, and the marine fauna is more varied, including a land crab (probably engraved after the White watercolour). The hermit crabs have disappeared, but the horseshoe crabs sport far more formidable claws. The fish are less recognizable and have been joined by some eels or water

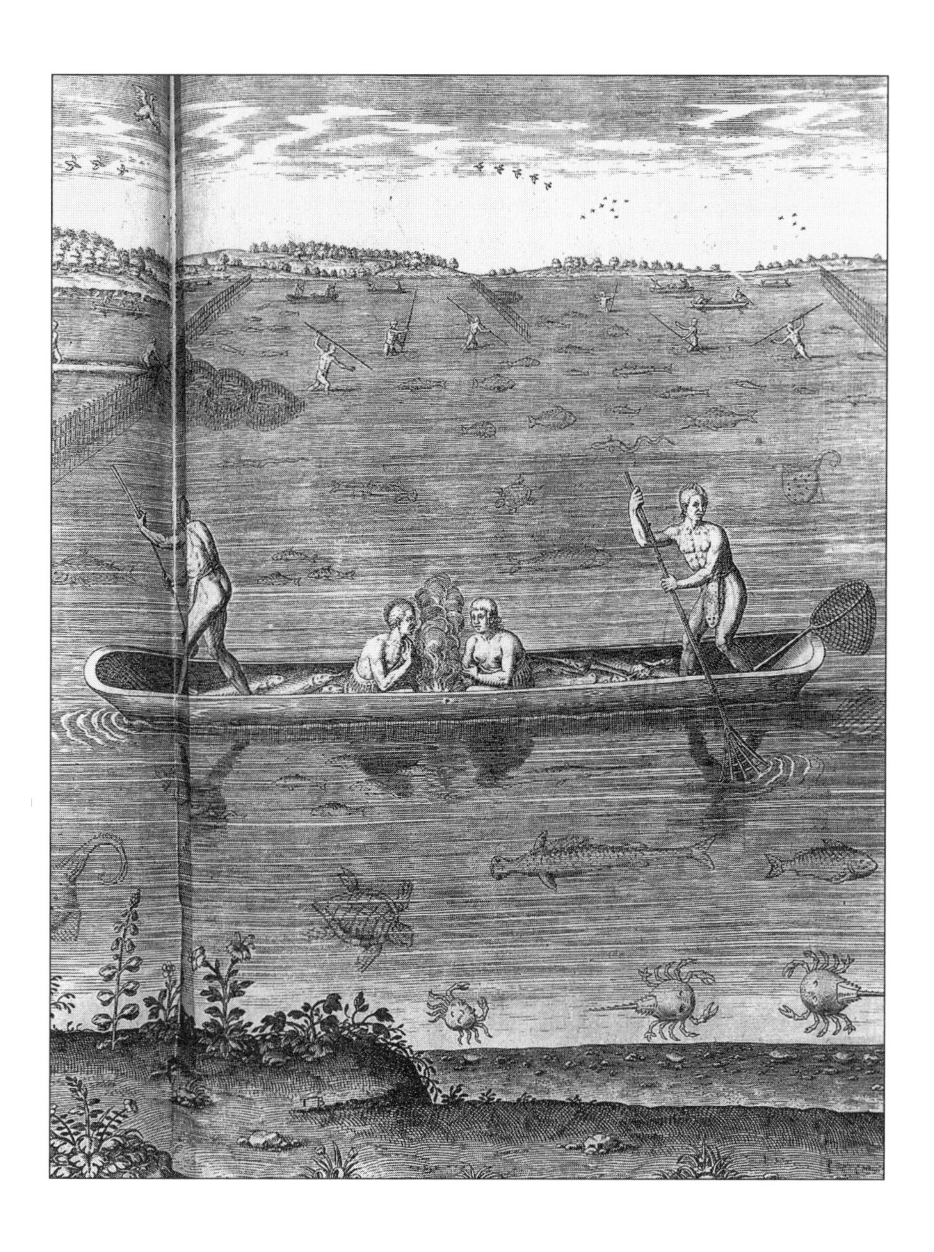

PLATE 9: Theodor de Bry (1528–1598), after John White (fl. 1577–1593), ***Their manner of fishynge in Virginia***, 1590. Copper engraving. Plate XIII from the German edition of Thomas Harriot, *A true and briefe discourse of the new found land of Virginia*, volume 1 in the *America* series or *Les grands voyages*, published by Theodor de Bry (Franckfurt am Mayn, 1590).

snakes, as well as by a number of turtles which appear to have been copied from White's drawing of the loggerhead turtle. De Bry has also added to the *cannow*-load a sturgeon, or perhaps a gar-pike, examples of which appear in a number of other engravings, poking out of baskets or roasting on grills. The birds have become generic ducks. The flowers in the foreground are more detailed but less well-realized, in that they seem to represent 'posies' rather than actual plants. In the watercolour, White shows clearly a tall spike of red flowers in the left foreground. This might be an early depiction of the cardinal flower, a spectacularly scarlet spike native to North America that grows in wet places. Thus, even when taking great pains to cut the image in copper, de Bry has rethought the original pictorial statement. The sea now teems with strange fish, and the peculiar-looking ray, which is barely seen in the original, is prominent in the engraving and very spotted. While White's disposition of the fish and invertebrates could not be said to be natural, in that the fish are 'posed' on the water, he has managed to convey a sense of habitat which is entirely missing in de Bry's stylized and static reworking, in which birds follow each other in a line and horseshoe crabs face off like mirror images.

What has de Bry done in this translation and why has he done it? The central image of the Native people in the canoe has remained much the same except that de Bry has replaced one of the middle figures with a female. Did women go fishing, or was this in the masculine sphere? He has also added many another canoes, and additional standing figures spearing fish. Did large parties go fishing or was de Bry concerned only with filling the visual field? The fish weir has been rendered much more complex, and a number of other weirs have been added. Is this American practice or European? The waters are now crowded with fish. Since we know that White was attempting to pose a number of animals together in one composite view, was de Bry simply trying to fit as many different kinds of things into the picture?[53] The horseshoe crabs and hammerhead sharks certainly count as curiosities, but Europeans were familiar with rays and crabs. One explanation for de Bry's exaggeration of the claws on the horseshoe crab may be to explicate the text, which refers to them as 'a certaine fishe like to a sea crabb.' Why have the birds become less distinct, while the fish have become more differentiated? Perhaps while, for White, given his interest in natural history, the birds were individual types, for de Bry they might have been just background. Thus, de Bry's engraving, though superficially like the original White watercolour, conveys both more and less information. It provides more variety of things, but it places them in a composition which has less veracity than the original. Bernadette Bucher, in her structural analysis of de Bry's work, terms this kind of assemblage a 'bricolage': 'If, like the *bricoleur*, the engravers draw their material here and there from a repertoire of forms that they use and transform to suit their circumstance or mood, the result of this *bricolage*, the rearrangement of those heteroclite forms into a visual narration of America, perhaps answers to an internal logic, as does myth.'[54] What Bucher suggests in her study is that de Bry read into the images of America both a golden age and the persecution of the Protestants. Just as he transformed the Native people into idealized versions of themselves, so, too, he transformed the landscape into a vaguely European vision, its difference marked only by the presence of the emblematic animals that identified the landscape as 'new.'[55]

The tendency of the engraver to be a *bricoleur* is accentuated both by the economic considerations of the printer – it was cheaper to copy or to buy a plate than to hire an artist to create one – and also by the new relationship of text and words. The development of printing led to a rupture in the web of meanings which had surrounded text and image in the hand-copied manuscript. This is particularly apparent when the reader examines the intricate patterning of letters and images in copies of prayer books and missals with marginal decorations and embellished initials. In printing, the separation between the block and the type is not only one of kind; it is physical. The cut block exists on its own, has meaning on its own, disentangled from the text which explains or describes it. 'When the graceful lines that linked text to marginal decoration were severed, pictures and words were disconnected. The former even while being reproduced were removed from their initial context and became more liable to being used indiscriminately. Relationships between text and illustration, verbal description and image, were subject to complex transpositions and disruptions.'[56] This is most obvious in the very early editions of printed books, such as Columbus's letters, as we noted in chapter 1. Here the illustrations borrowed from a book about Mediterranean harbours could be reused and slightly modified to suggest the idea of discovery and the idea of the Native people. Even at a later period, Dürer's rhinoceros could serve both as a representation of the King of Portugal's animal and, without accompanying text, as the true representation of an animal recently shipped from Asia over two centuries later. Thevet, as we noted in chapter 1, could relocate the su in Florida from its original home in Patagonia, by disassociating it from the text in his first book and from its context in Magellan's voyage, and providing it with a new text that authenticated it in a new location. Lest it be assumed that this disconnection between text and image occurred only rarely and in connection with the exotic, Francis Haskell's book *History and Its Images* lists a vast number of instances when the printed image is dissociated from its original context and used not only in an inappropriate manner, but often in a way which we would today consider dishonest. In his discussion of the vogue for collections of historical portraits in the seventeenth century, he writes that, despite the claims of the publishers,

> ... it is by no means simple to verify the accuracy of what was claimed to be true likeness, and opportunities for fraud were plentiful – and could be seized with great recklessness. Thus in the 1640s the London print-publisher Peter Stent arranged for a copy to be made with an etching by Rembrandt dating from only a few years earlier (believed to represent his father) and used it as a portrait of Thomas More, despite the fact that the features bore not even the faintest resemblance to those of More which were familiar enough through many versions of Holbein's portraits and even through medals.

He describes as well the creation of a whole series of true 'portraits' of the fifth-century kings of France, for which no contemporary medals or likenesses exist, and notes that it is not 'clear whether the images they produced were actually intended to deceive rather than to evoke a general impression (based on literary sources) of an unknown world.'[57]

There is another aspect to the relationship between text and image that is evident in this early period. There appears to exist a distrust of images as true representatives of things in themselves. There remains a sense that the word itself conveys the 'essence' of the thing. Gesner, Aldrovandi, and other authors of the sixteenth century spend considerable time on the analysis of names, not only because, lacking a common scientific nomenclature, they must record as many names as possible in as many different languages to ensure that their readers will recognize the animal, but also because they retain the medieval belief in the power of the name as a sign of the thing itself. Thus, when confronted with an unfamiliar animal, the artist returns to the name or the textual description. In the *Hortus Sanitatis* of 1491, which included the *Tractatus de Avibus*, the artist has pictured what must be from the description a ptarmigan, as a bird with the head of a hare. The Latin name is 'lagophus,' which refers to the ptarmigan's feathered feet, which resemble those of a rabbit or hare. In the same work is a bird with two horsehoes in its beak. This is the ostrich, which Pliny described as being able to digest iron. We can speculate that, in the absence of detailed information on the rhinoceros, even Dürer took his clue about the overall appearance of the hide from his correspondent's description of the beast as having a 'thick shell.' This examination of the translation of original image into the 'repeated pictorial statement' provides us with another gloss on the role of images and their relationship to knowledge in the early modern period. Pictorial rendering and textual description existed in an uneasy relationship. While the original manuscript image might stand for the thing itself, its reproduction, mediated through the printing process, proved problematic as a conveyor of information. The next chapter examines the nature of verisimilitude from another perspective – the transformation of the image and its translation into an accurate rendering of the thing itself into the 'scientific illustration.' It is in the botanical literature, and particularly in the transition from herbals to florilegia, that this process can be most clearly seen.

three

THE LIVING IMAGE

Samuel de Champlain (*ca* 1567–1635) first sailed to Canada in 1603 as an observer on a French colonizing expedition. In 1608 he returned to Canada and founded Quebec, making la Nouvelle France his second home. In 1613, he published an account of his explorations in the New World: *Les voyages du Sieur de Champlain Xaintongeois, Capitaine ordinaire pour le Roy*, subtitled *Iournal tres-fidele des observations enrichi de quantité de figures* ... While illustrations based on Champlain's detailed sketches of Native life were included in his 1620 *Voyages et descouvertures*, the 1613 volume was enriched primarily with small maps of the areas explored, as well as with one large map of eastern North America, including Newfoundland and Labrador, Acadia, the St Lawrence and beginning of the Great Lakes, and part of New England. W.F. Ganong praises Champlain for his abilities as a cartographer, but obvious as well are his skills as an artist. The small maps show, in the manner of sixteenth-century cartographic style, tiny scattered drawings of wildlife – a stylized deer and rabbit, a few ducks, a fish – which might have been added by the engraver. The large-scale map folded into the back cover of the book is exceptional, however, for the detailed renderings of Native people, plants, and animals (plate 10). The animals are figured on both sea and land, and Champlain was obviously a curious and close observer. Since many of the images are unique, it has been assumed that they are from Champlain's sketches, though the map was prepared by David Pelletier. In the sea, then, Champlain has figured the *bar* (Ganong: striped bass), *molue* (Ganong: cod), *gros chabos* (Ganong: sculpin), *lou marin* (Ganong: seal); *chien de mer* (Ganong: dogfish), *siguenoc*

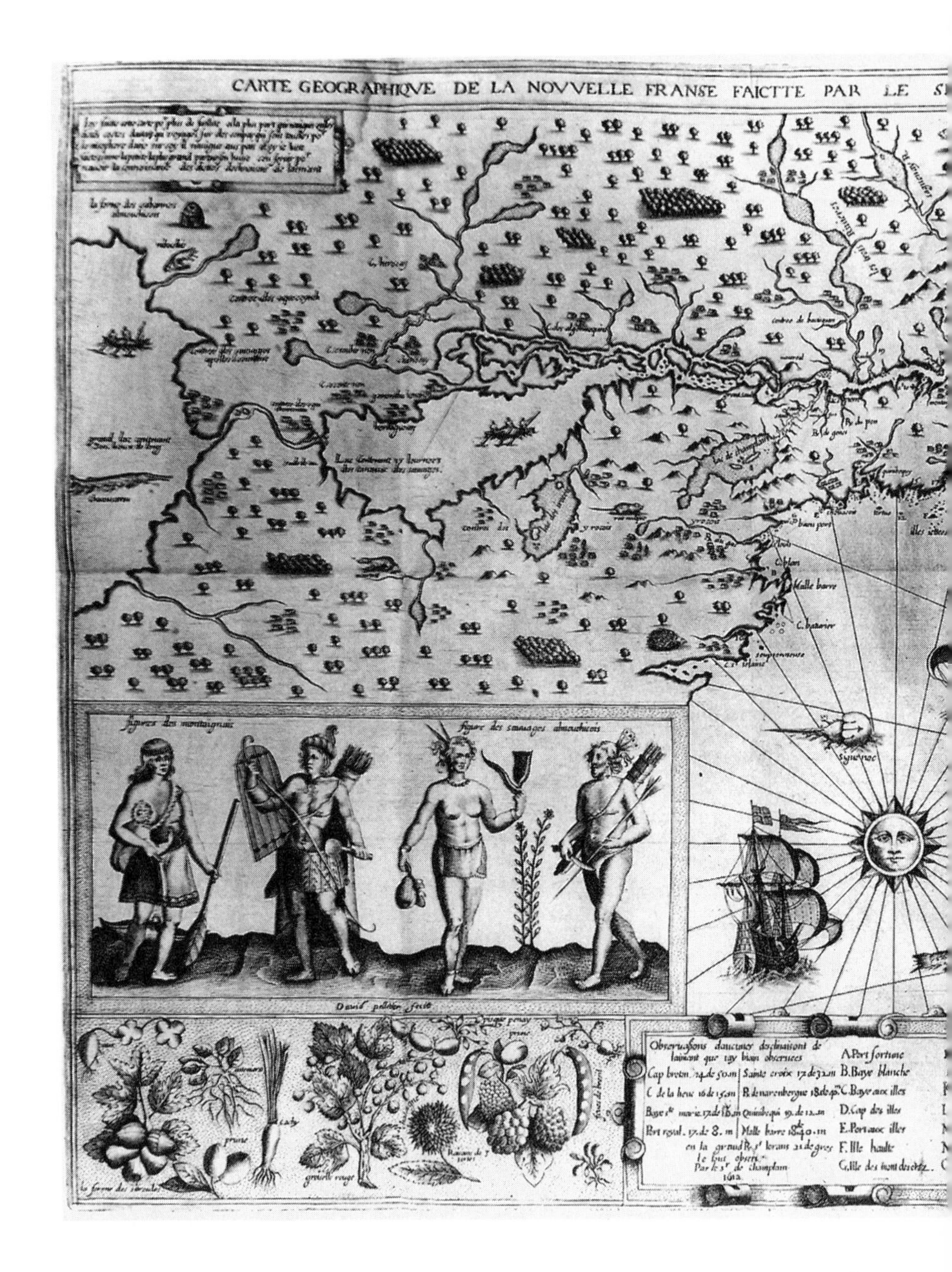

PLATE 10: David Pelletier, after Samuel de Champlain (*ca* 1567–1635), ***Map of New France***, 1613. Engraving. From *Les voyages du Sieur de Champlain Xaintongeois*, subtitled *Iournal tres-fidele des observations enrichi de quantité de figures ...* (Paris, 1613). The large-scale map is folded into the back cover of the 1613 edition.

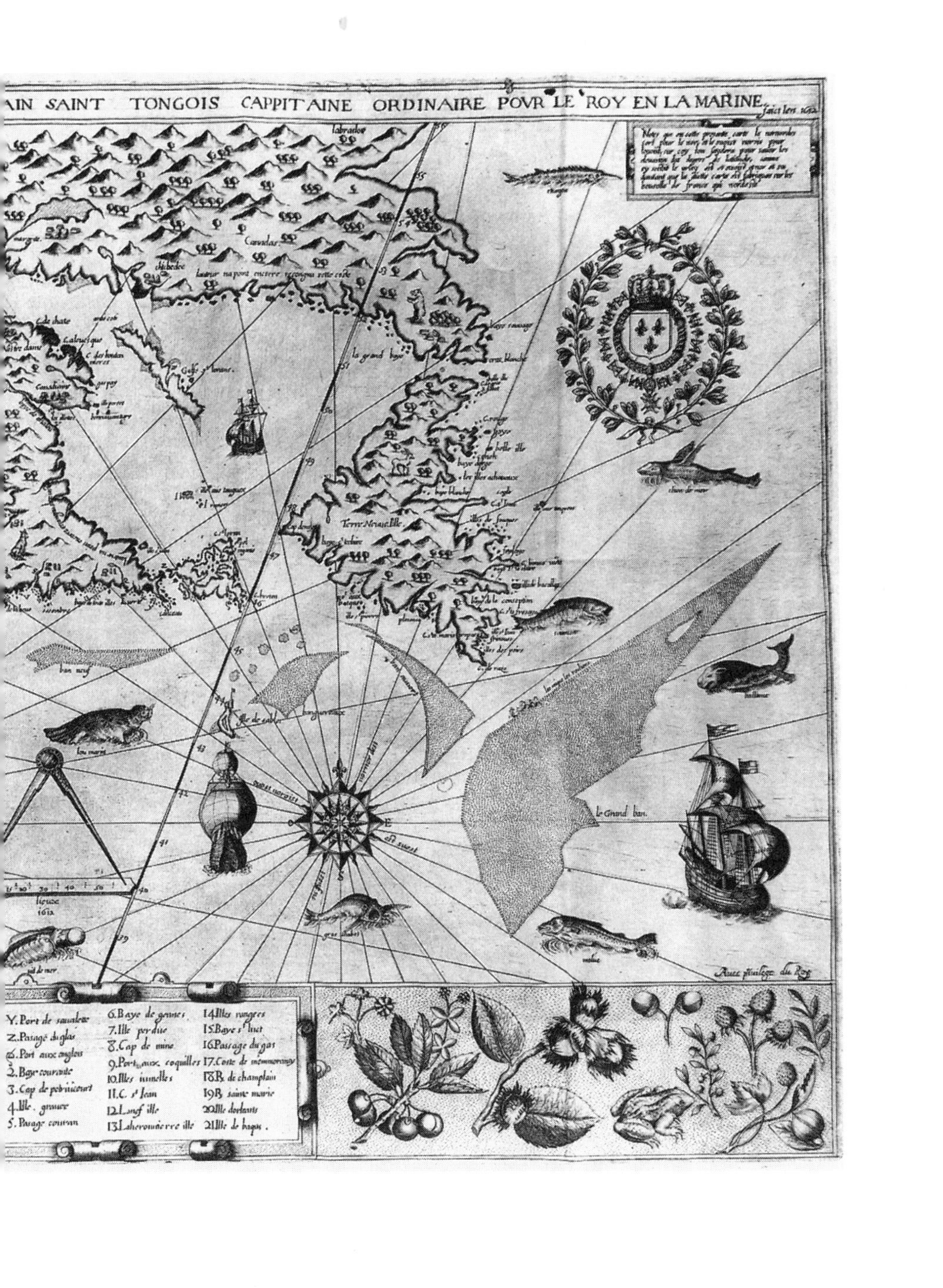
AIN SAINT TONGOIS CAPPITAINE ORDINAIRE POVR LE ROY EN LA MARINE
labrador
Canadas
Terre Neuue Ille
le grand baye
ban neuf
Ille de sable
le Grand ban
Auec priuilege du Roy
Y. Port de sauualettre
Z. Pasage du glas
6. Port aux anglois
2. Baye courante
3. Cap de potriucourt
4. Ille . grauee
5. Pasage couruan
6. Baye de gennes
7. Ille perdue
8. Cap de mine
9. Port aux coquilles
10. Illes iumelles
11. C. st Jean
12. Longe ille
13. Laheronnierre ille
14. Illes rangees
15. Baye st luct
16. Passage du gas
17. Coste de memmorancy
18. B. de champlain
19. B. sainte marie
20. Ille dorlans
21. Ille de baqus .

(Ganong: horseshoe crab), *ballame* (Ganong: ballein/Greenland whale), *eturgon* (Ganong: sturgeon), *scumisn* (according to Ganong, probably *saumon*, or salmon), and finally the *vit* or *uit de mer* (Ganong cannot identify this 'mollusc,' but it appears to resemble a sea cucumber, a number of species of which are common around the Bay of Fundy area).[1]

On the land appear the usual set of diagnostic northern-hemisphere animals, with some interesting additions. To the bear, deer, wolf, and porcupine, Champlain has added the *castor* (beaver), the *martre* (Ganong: pine marten), *nibachis* (Ganong: a probable M'kmaq word for 'raccoon,' though the animal is placed north of the Great Lakes), and *rat musque* (muskrat). Afloat in the waters of the Great Lakes is another fish, the *chaousarou*, not identified by Ganong, but from its appearance it can be none other than a garpike, common in the Great Lakes, and certainly a curiosity, having hard scales like the sturgeon, and a formidable set of 'alligator' jaws with sharp teeth. In the seventy years since Desceliers and Gastaldi had drawn their maps, the known fauna has increased, and the depictions of very common animals such as raccoons, beavers, and muskrats are recognizable and well rendered. Where Cartier had commented on their abundance, he was unable to provide the original images which might have allowed engravers and map-makers to include them on their maps, and, perhaps more significantly, they were not part of the standard Nearctic/Palaearctic map fauna that identified the north. Champlain's animals are the animals of a resident. He would have had an opportunity to see them in the wild and certainly to examine their skins, which were objects of trade. Interestingly, none of these animals (with the exception of a group of deer) appears on the 1632 map *Carte de la Nouvelle France*. Swift's remark on the use of animals for want of towns perhaps has some justice, since the 1632 map is dense with Native settlements and shows the inland watercourses in some detail (plate 11). The sea in this map is still, however, populated with fish and whales.

The 1613 map also includes a vignette of four Native people, 'figures des montaignais' and 'figures des sauvages almouchicois,' one of whom holds a ripe ear of corn and a gourd. Like the de Bry renderings of Native people, the figures are classicized, though the details of clothing and accessories are distinctive. What distinguishes the Champlain map, however, is the inclusion of a border containing well-rendered plants. While early explorers like Cartier had certainly noted the abundance of trees and fruits (see above), in contradis-

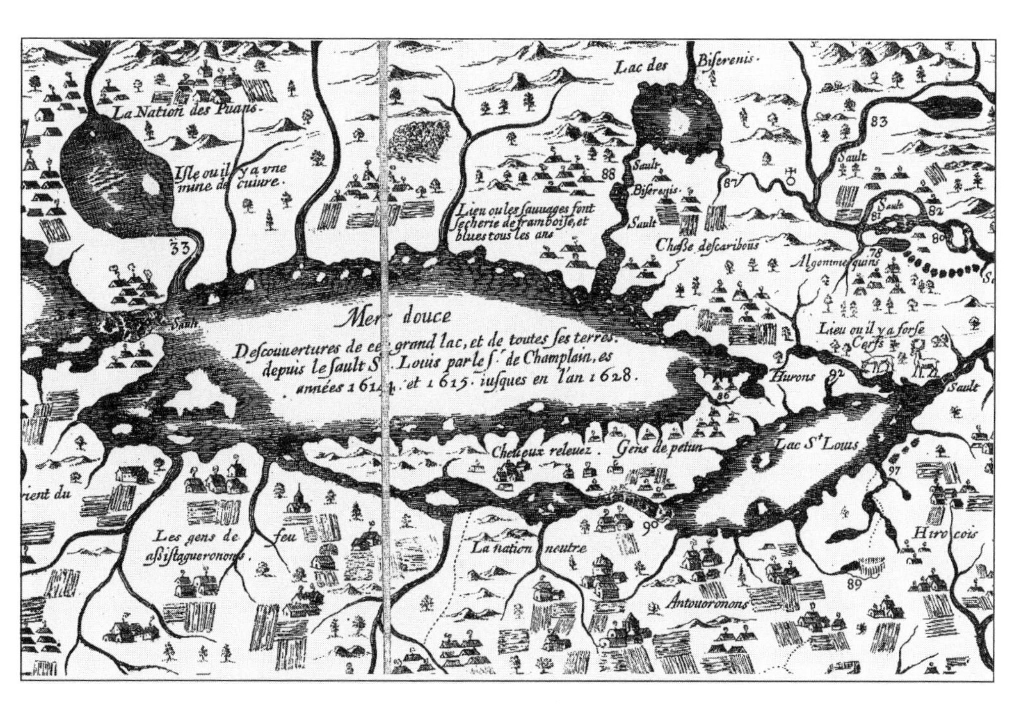

PLATE 11: After Samuel de Champlain (*ca* 1567–1635), ***Map of New France***, 1632. Engraving. From *Les Voyages de la Nouvelle France occidentale, dicte Canada: Faits par le Sr. de Champlain* ... (Paris, 1632).

tinction to the animals, they had not been depicted, except schematically on maps. Both John White and Jacques le Moyne de Morgues, skilled botanical artists, had included stylized grapevines on map views included in de Bry (plate 12), and their depictions of landscape showed some diversity of trees, such as cedars, palms, and deciduous species, but with the exception of crop plants (*pompions*, or pumpkins, corn, sunflowers, and tobacco) and some recognizable milkweed and cattails included in engravings of le Moyne's work (plate 13), few sixteenth-century illustrations of North American landscape revealed very much of the vegetative abundance which so struck the first explorers. Champlain's engraver has divided the border into three parts: on the left, a number of text-identified plants; in the middle, a legend to the map; and, on the right, more plants not identified by text. The plants on the left include two types of *prunes* (plums), *raisains* [*sic*] *de 3 sortes* (grapes), *groiselle rouge* (red currant), *feves de bresil* (likely the wild bean, apparently much

PLATE 12: Theodor de Bry (1528–1598), after Jacques le Moyne de Morgues (d. 1588), Map of Portus Regalis, 1591. Copper engraving. Plate V from volume 2 in the *America* series or *Les grands voyages*, published by Theodor de Bry (Frankfurt am Main, 1591).

prized by the Native peoples),[2] the *chataigne* (American sweet chestnut), *sitroulos* (*citrouille*, squash or gourds), and a number of roots, only one of which, the *pisque penay*, is identified in Ganong (ground nut, *Apios tuberosa*).[3] Edible roots were a staple of the Native diet and it is likely that the four plants depicted with their roots were prized for that part. Ganong has been unable to identify the *cachy*, though its leaves resemble those of the lily or rush families, various members of which were eaten by the Amerindians. The *aux* is likewise unidentified and, in distinction to the other plants, its leaves or flowers are poorly drawn, though it may well be the wild garlic.[4] The *astemara* may resemble in name and configuration the wild ginger (asarabacca), another plant whose root was used by the Native people.[5] The plants on the right side of the border are not identified but include a

PLATE 13: Theodor de Bry (1528–1598), after Jacques le Moyne de Morgues (d. 1588), ***Apalatcy***, 1591. Copper engraving. Plate XLI from volume 2 in the *America* series or *Les grands voyages*, published by Theodor de Bry (Frankfurt am Main, 1591).

cherry, strawberry, walnut, acorns, and perhaps a blueberry, which Champlain relates sustained them on their journey along the French River, as well as an anomalous frog.

Champlain's is not the only map of the period to feature plant life. Marc Lescarbot's 1611 map, while less detailed than Champlain's, includes a kind of cartouche in the bottom left-hand corner showing wild grapes, ears of ripe corn, and three rows of young corn plants (plate 14). By 1600 the vegetative products of the New World were becoming, like New World animals, more common in Europe and more often depicted. We have already noted Ligozzi's very fine watercolour drawings (*ca* 1580) of a pineapple and the 'Marvel of Peru,' but what made it possible in 1613 for Champlain and his engraver to depict with delicacy and a fair degree of realism a variety of unfamiliar plants

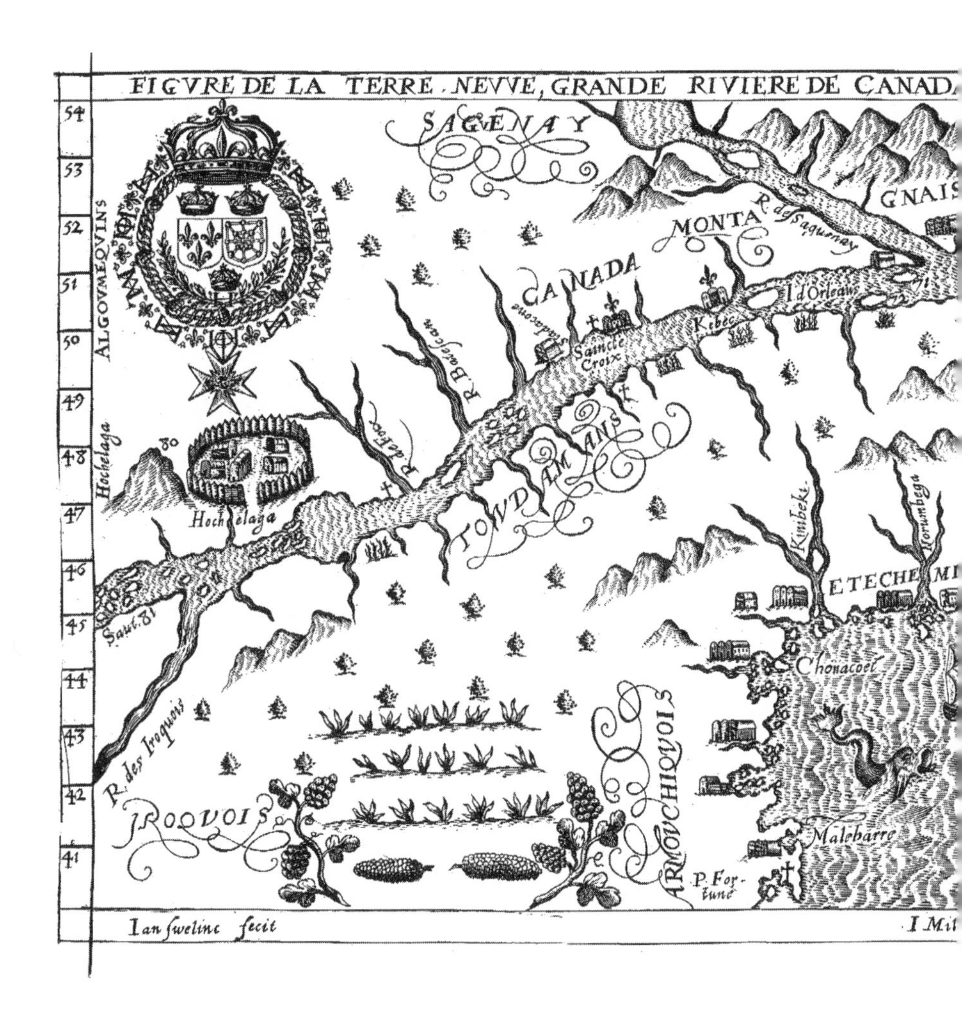

PLATE 14: Jan Swelinc (b. 1601?), after Marc Lescarbot (1570–1642), ***Figure de la Terre Neuve, Grande Riviere de Canada, et côtes de l'Ocean en la Nouvelle France***, 1611. Copper engraving.

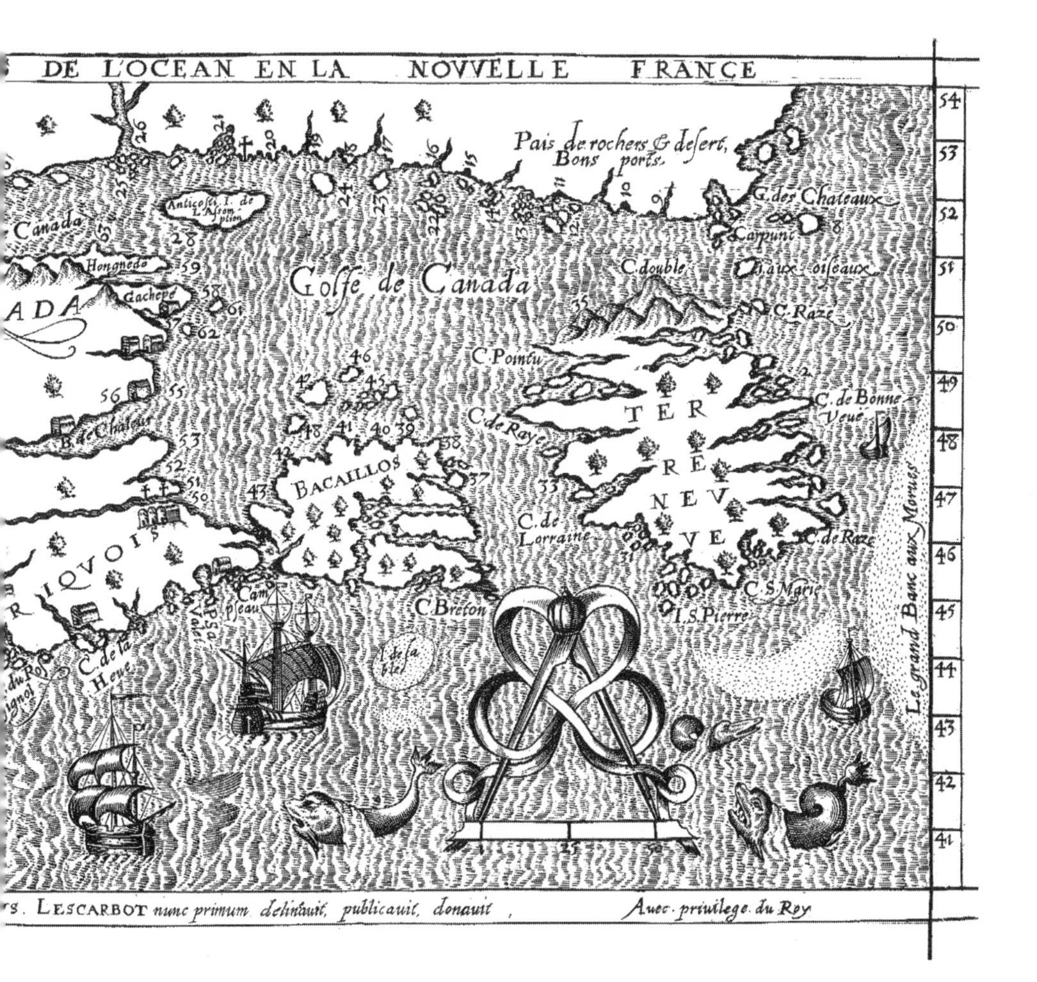
DE L'OCEAN EN LA NOUVELLE FRANCE
Pais de rochers & desert, Bons ports.
Golfe de Canada
Canada
Hongnedo
Gachepé
Anticosti I. de l'Assomption
G. des Chateaux
Carpunt
C. double
I. aux oiseaux
C. Raze
C. Pointu
C. de Bonne Veue
C. de Raye
BACAILLOS
TER RE NEV VE
C. de Lorraine
C. de Raze
C. S. Marie
I. S. Pierre
C. Breton
Campseau
B. de Chaleur
C. de la Heve
Le grand Banc aux Moruës
S. LESCARBOT nunc primum delineauit, publicauit, donauit
Auec priuilege du Roy

was the transformation in botanical literature and illustration that occurred from the mid-sixteenth to the late-seventeenth century. During that period, herbals became florilegia, and the study of simples became the study of plants. In the eighteenth century the study and classification of plants was to transform the way in which people viewed the natural world. An appreciation of the history and formalization of botany, in which the productions of the New World played no small part, is important to an understanding of this change in world-view. By looking closely at two books, one published in 1635 and the second in 1744, which illustrate some of the flora of the northern half of the New World, we can examine both the growth of botanical knowledge and its relationship to botanical illustration.

Cornut and the *Canadensium Plantarum*

Jacques-Philippe Cornut (*ca* 1606–1651) published his *Canadensium Plantarum aliarúmque nondum editarum Historia* ... in 1635 in Paris. Cornut was a medical doctor, and his book described for the first time about thirty North American plants, chiefly from the gardens of the Faculty of Medicine in Paris. Eighty-six plants are described in the book, most of them illustrated with a full-page copper engraving showing the root, stem, leaves, flowers, and sometimes fruit. In some plates, as in plate XXXVIII, *Apocynum minus rectum Canadense* (common milkweed), details of the unique seed case, the seeds with their filamentous pappi, and the flower head are added. In plate XL, featuring *Edera trifolia canadensis* (poison ivy), the fruit is enlarged and the manner in which the plant spreads is indicated by the horizontal root with stems emerging upwards (plate 15). The roots are included on almost all plates, a reflection of their importance in medicinal use (plate 16). The fact that the *Historia* was grounded in medical botany is important to understanding the convention by which Cornut and his artist worked in producing and engraving the illustrations.[6] Cornut's work stands near the end of a very old tradition of herbal literature, and the illustrations reflect a convention established in the mid-sixteenth century. At the same time, Cornut's illustrations were not printed from woodcuts, but from engraved copper plates, and his text reflected not only medical usage, but horticultural observations. The title of Cornut's work also suggests that he has at least attempted to prepare a regional flora, and he has prepared it by working with a living collection, a

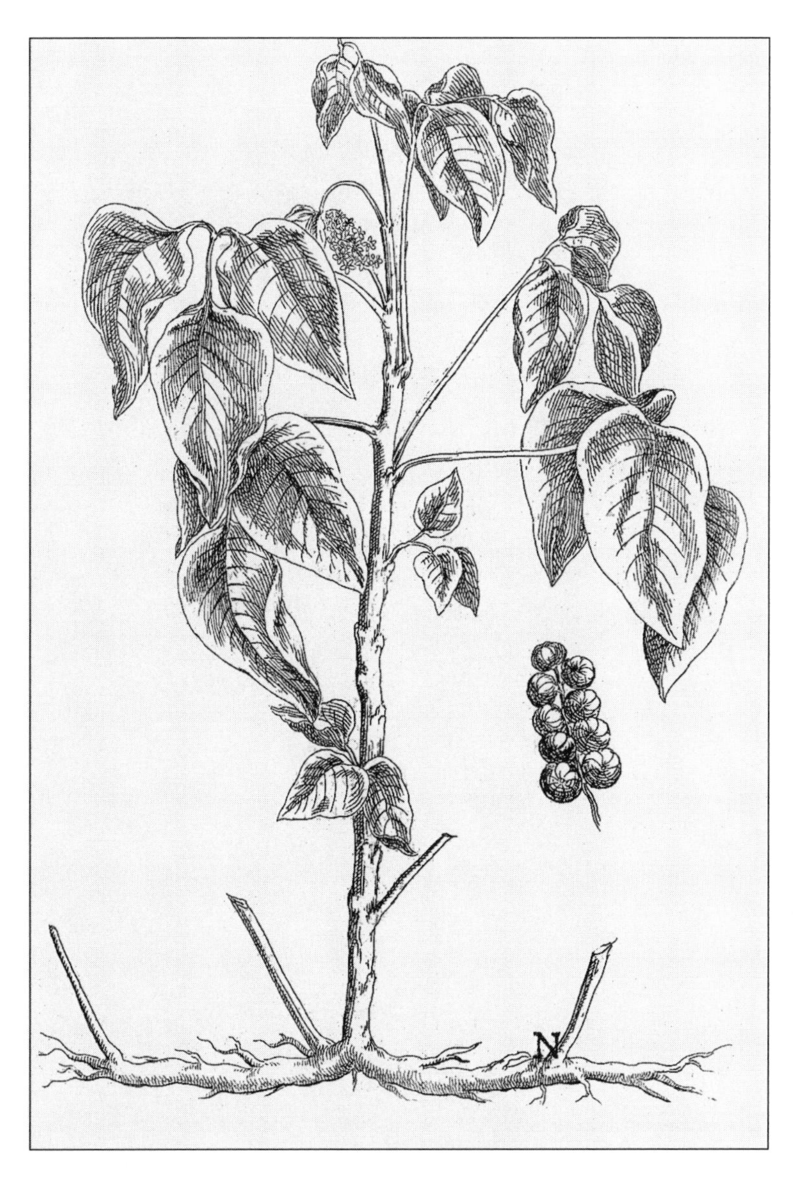

PLATE 15: ***Edera trifolia canadensis***, 1635. Engraving. From Jacques Philippe Cornut (*ca* 1606–1651), *Canadensium Plantarum aliarúmque nondum editarum Historia* ... (Paris, 1635), plate XL, 97. Poison ivy (*Toxicodendron radicans*) continues to be a scourge for travellers in the woods of eastern Canada. The crushed root was, however, applied by the Native peoples as a poultice for skin disease, and it is also reported that an eighteenth-century French physician used the plant in the treatment of skin infection.

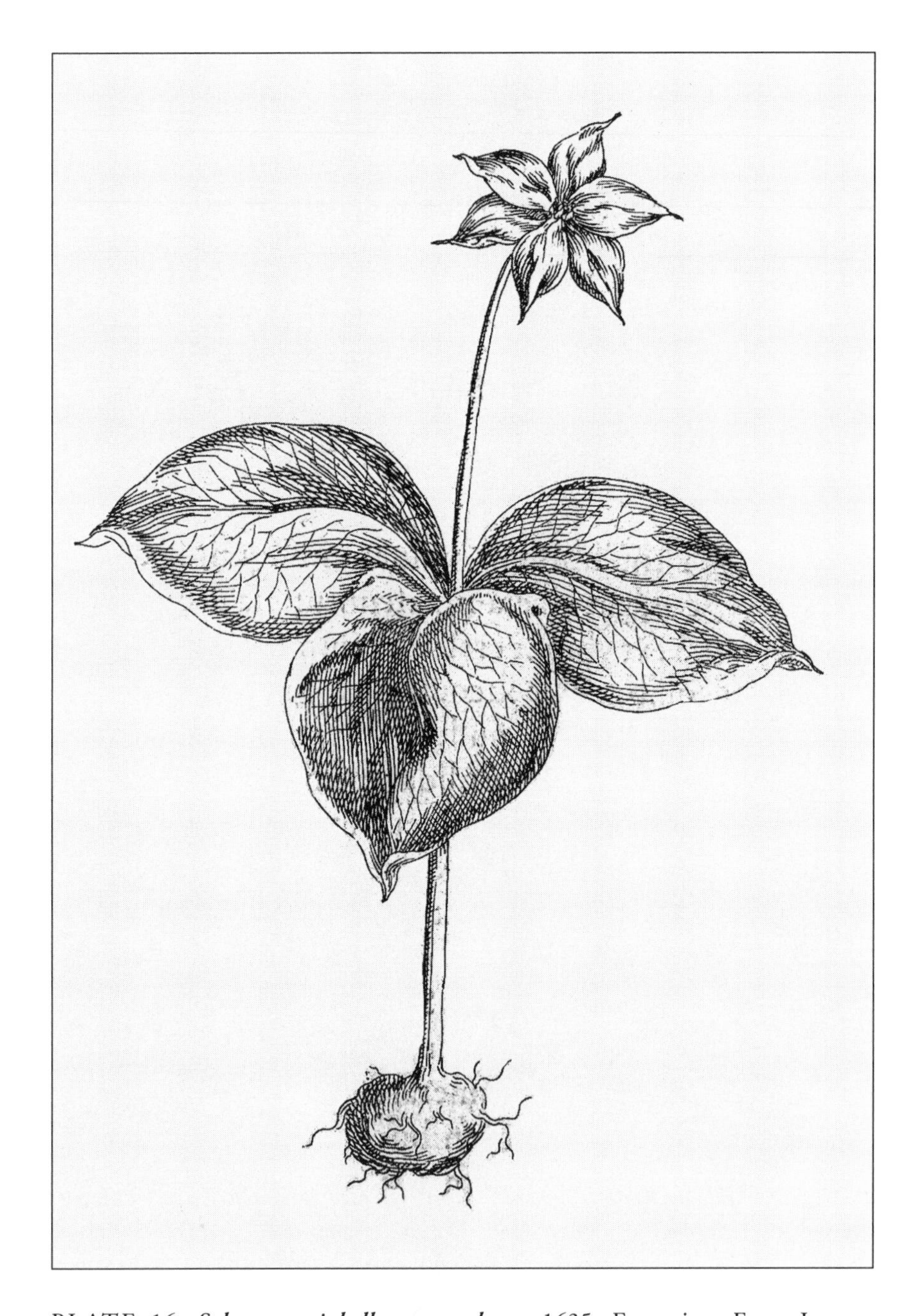

PLATE 16: ***Solanum triphyllum canadense***, 1635. Engraving. From Jacques Philippe Cornut (*ca* 1606–1651), *Canadensium Plantarum aliarúmque nondum editarum Historia* ... (Paris, 1635), 167.

garden. These two aspects of Cornut's work, herbal on the one hand and regional flora on the other, mark it as transitional, between the Renaissance and the early modern period. A close reading of the illustrations and their conventions provides insights not only into Cornut's work, but into the use of visual information at the beginning of a period in which the role of the image in botany was to become paramount.

The Herbal Tradition

The herbal illustrated a collection of plants considered chiefly in terms of their medicinal properties. Prior to the interest in chemical remedies that begins in the sixteenth century, almost all medicines were derived from plant materials. Doctors relied for their prescriptions on the classical, that is, Greek and Roman, pharmacopoeia, as it had been passed down, often imperfectly, through the medium of manuscript herbals. For the most part the herbals were based on *De materia medica* by the Greek physician Dioscorides, written in the first century A.D. Much botanical writing in the medieval period was based on Dioscorides' text, and attempts to identify plants and their cures relied on his brief descriptions and the corrupted, much-copied illustrations. Cornut obviously had his Dioscorides as well as his Pliny and a number of other classical and contemporary authors at his side and found it appropriate to compare a particular plant with a description of Dioscorides', even though he acknowledged that, in the Greek physician's time, neither Canada nor America had been discovered.[7] It is easy to dismiss the search for authority in a 1,500–year-old source, but, before the application of a scientific method to pharmaceutical trials, the action of a particular plant could be determined only through patient observation or transmitted information.[8] The respect of Renaissance scholars for classical learning increased their confidence in these remedies of the ancients. Karen Reeds notes that, 'by the 1530s, then, Theophrastus, Pliny, Dioscorides and Galen were all available in up-to-date printed editions, and the Greek treatises had been freshly translated into Latin. The large number of editions and translations is a good index of the popularity of these authors and the respect in which their work was held.'[9] At the same time, however, as the printed books reinforced traditional knowledge, they also also pointed out its deficiencies, both in the practice of European physicians and in light of discoveries in the New World.

If the classical authors reiterated one point, it was the importance of fieldwork: 'Prefaces to Renaissance herbals regularly quoted the passages that showed Galen travelling to Palestine ..., Dioscorides watching herbs sprout, grow, flower, fruit and die, and Pliny's friend, Antonius Castor, studying plants in his garden even after reaching his hundredth year.'[10] Renaissance botanists began to look with new eyes on the plants of field and garden, and as they did they realized a number of things. The first, and any glance at many late-fifteenth-century printed herbals will confirm the observation, was that accurate images of plants drawn from life would greatly help in identification. Plants, like the animals in the *Physiologus* and fables, had suffered from uninformed hand-copying. When the manuscript herbals were first printed, the woodcutters followed the illustrations of the originals, reproducing schematic icons of particular plants. Ford suggests that the use of unnatural icons as reference images might be, not a sign of incompetence, but rather an attempt to preserve the arcane nature of the knowledge,[11] whereas Boas feels that the cutters simply copied the originals faithfully because they saw them as illustrating, not nature, but the text.[12] In an oral and manuscript tradition, as Eisenstein has pointed out, the reliance is on neither word nor image. The teaching is passed on from master to student, and in this case it is possible to imagine even the very corrupted images of the early printed herbals serving as *aides-mémoires*. Arber's conclusion adds some confirmation to this view:

> We may safely conclude that the draughtsman knew quite well ... that he was not representing the plant as it was, and that he intentionally gave a conventional rendering, which did not profess to be more than an indication of certain distinctive features ... For instance, a plant such as the houseleek may be represented growing on the roof of a house – the plant being about three times the size of the building. It is evident that the little house was introduced merely to convey graphic information as to the habitat of the plant concerned, and that the scale on which it was conceived was simply a matter of convenience. Before an art can be appreciated, its conventions must be accepted. It would be absurd to quarrel with the illustrations we have just described ... as to condemn grand opera because, in real life, men and women do not converse in song.[13]

There also existed classical antecedents for a reticence about reliance on the use of image as the bearer of accurate information. Pliny, while praising the 'attractive method' of some Greek botanists, commented on the difficulty of portraying plants:

> the subject had been treated by Greek writers, whom we have mentioned in their proper places: of these Crateuas, Dionysius and Metrodorus adopted a most attractive method, though one which makes clear little else except the difficulty of employing it. For they painted likenesses of the plants and then wrote under them their properties. But not only is a picture misleading when the colours are so many, particularly as the aim is to copy Nature, but besides this, much imperfection arises from the manifold hazards in the accuracy of copyists. In addition, it is not enough for each plant to be painted at one period only of its life, since it alters its appearance with the fourfold changes of the year.[14]

Part of Pliny's criticism is attributable to the manuscript tradition, in which the original image degenerates with each copy. Eisenstein and Ivins have pointed out that the technology of printing 'repeated pictorial statements' allowed the repetition of images without degeneration, but the advantages of this new accuracy of reproduction were not immediately appreciated by authors and printers. In 1500, Hieronymus Braunschweig could still conclude his work on distillation by warning readers that they must pay attention to the text, since the woodcut figures had been included as 'nothing more than a feast for the eyes, and for the information of those who cannot read or write.'[15] Nevertheless, Renaissance botanists did realize a need to improve both information and illustration. The proliferation of printed herbals meant that people began to rely on the book as a 'silent teacher.' It was no longer enough that the illustration serve to remind the reader of a plant already learned. The illustration should help to identify the plant collected in the field, and by 1530 readers refused to rely on text alone. In that year Otto Brunfels (1488–1534) published a herbal entitled *Herbarum Vivae Eicones*, in which the illustrations were painted, not from existing 'reference' images, but from life.

It was not simply the new technology, however, that produced this new kind of herbal; it was also a new kind of sensibility regarding the objects of the natural world. The transformation from patterning or caricature to naturalistic representation was embodied in the work of the illustrators for the new naturalists. Brunfels employed Dürer's pupil Hans Weiditz, and it is the sensibility of Dürer's 'Columbine' or 'Great Piece of Turf' that informs the work of Weiditz. The watercolour image of a plant or animal had become accepted as a medium for the exchange of information among sixteenth-century naturalists, and an image 'drawn from life,' and printed with sufficient resolution, allowed them to spread further their discoveries and to communicate vital information about plant identification to medical practitioners. For Brunfels's *Herbal*, Weiditz prepared careful coloured drawings in the manner of his master, recording the wilted leaf, the insect damage. Brunfels included a Latin poem in which he compared the work of Wieditz to that of 'Clarus Appelles,' the Greek artist who, Pliny asserted, painted a horse so lifelike that real horses neighed at it. Weiditz's drawings were widely acknowledged as superior to Brunfels's text, but were criticized for being too much portraits, and too difficult to identify with plants in the field. Resistance to using images as information was still common even after the publication of Brunfels's *Herbal*, and Hieronymus Bock (1498–1554) refused to have illustrations in the first 1533 edition of his herbal, *New Kreüter Buch*, fearing that they would distract the reader from the text descriptions, so clear and expressive that even Konrad Gesner, a great collector of drawings, acknowledged that no painter could describe the plant better.[16] Bock's was essentially a rearguard action, likely born out of concern for the problems of translation into woodcuts as well as a Protestant objection to the use of the image over the word. In later editions, however, he changed his mind, for reasons to be discussed below, and used some of the blocks cut for Leonhard Fuchs's (1501–1566) masterpiece *De Historia Stirpium* (1542).

Fuchs's work has been described by Charles Singer as 'the highwater mark of the Renaissance herbal.'[17] *De Historia Stirpium* was conceived by Fuchs from the beginning as an integrated work in which text and illustration would complement each other. Fuchs understood the use of accurate illustration and noted in his preface that 'it is the case with many plants that no words can describe them so they can be recognized. If, however, they are held before the eyes in a picture, then they are understood immediately at first glance.'[18] In an effort to ensure that the illustrations were 'understood immediately,' Fuchs hired the best artists and cutters and guided the entire work from painting to drawing on the block, cutting, and printing. He also attempted to remedy both the criticisms of Pliny and the complaint of portraiture that had been levelled at Weiditz:

> As far as concerns the pictures themselves, each of which is positively delineated according to the features and likeness of the living plants, we have taken peculiar care that they should be most perfect; and moreover, we have devoted the greatest diligence to secure that every plant should be depicted with its own roots, stalks, leaves, flowers, seeds and fruits. Furthermore, we have purposely and deliberately avoided the obliteration of the natural form of the plants by shadows, and other less necessary things, by which the delineators sometimes try to win artistic glory: and we have not allowed the craftsmen so to indulge their whims as to cause the drawing not to correspond accurately to the truth ...[19]

Fuchs's illustrations were so successful that a number of them were used not only by Bock, but also by Rembert Dodoens (1517–1585) in his Flemish herbal of 1554 and by Gerard in 1597, and survived to be reprinted in a botanical book published in Zurich in 1774. Dodoens's later herbals were published by Christophe Plantin of Antwerp, who amassed a large number of woodcuts printed in the works of de l'Obel and de l'Écluse. This collection of woodcuts became extremely popular and was used in herbals well into the seventeenth century. Arber points out that 'the woodcut of a clematis, which was first seen in Dodoens's *Pemptades* of 1583, reappears either in identical form, or more or less accurately copied in works by de l'Obel, de l'Écluse, Gerard, Parkinson, Jean Bauhin, Chabraeus, and Petiver,' being used for the last time in Johnson's 1636 edition of Gerard's *Herball*.[20] Thus, the conventions of mid-sixteenth-century herbal illustration were transmitted more or less intact to mid-seventeenth-century treatises. By the time of Cornut's publication, the inclusion of illustrations was not only accepted, but expected. Cornut did not use woodcuts, but copper engravings, and, as Arber points out, the fact that the copperplate engravings were printed separately from text contributed in the first half of the seventeenth century to the development of the picture book, or florilegium: 'The consequence is that books with fine engravings appeal rather to the wealthy amateur, whose influence is patent in the seventeenth-century botanical books, in which the theme shows a definite shift towards horticulture.'[21] We will look now to the sources for Cornut's illustrations and the history of the depiction of the vegetable productions of the New World.

The Sea of Simples

Most of our discussion has centred around European herbals based on European plants. The realization by the herbalists that the Dioscorides text referred to plants of the Mediterranean littoral, not to those of northern Europe, led first to field trips to examine the classic flora, like that of the author of the *Gart der Gesundheit*, published at Mainz in 1485. In the preface the author notes that, in publishing the herbal, he

> ... came to the conclusion that I could not perform any more honourable, useful or holy work or labour than to compile a book in which should be contained the virtue and nature of many herbs and other created things, together with their true colours and form, for the help of all the world and the common good ... But, when in the process of the work, I turned to the drawing and depicting of the herbs, I marked that there are many precious herbs which do not grow here in these German lands, so that I could not draw them with their true colours and forms, except from hearsay. Therefore I left the work unfinished that I had begun, and laid aside my pen ... I took with me a painter ready of wit, and cunning and subtle of hand. And so we journeyed from Germany through Italy, ... Greece, Corfu, Morea, Candia, Rhodes and Cyprus to the Promised Land ... and thence through Arabia Minor to Mount Sinai, ... and also Alexandria in Egypt ... In wandering through these kingdoms and lands, I diligently sought after the herbs there, and had them depicted and drawn, with their true colour and form ... And this book is called in Latin, *Ortus Sanitatis*, and in German *gart d'gesuntheyt*. In this garden are to be found the power and virtues of 435 plants and other created things, which serve for the health of man, and are commonly used in apothecaries' shops for medicine.[22]

While the author of the *Gart* desired to picture the herbs of Dioscorides, other authors began to include in their herbals plants with proven medical efficacy, probably learned from the local herbwomen, that did not feature in the classical pharmacopoeia. Brunfels and Weiditz included in their works plants called '*herbae nudae*,' which were not mentioned by Dioscorides. The use of the term is interesting, since it points to the importance of textual description. The herbs were 'bare,' not because they were not of use, but because they had no ascriptions, no history. The text of the classical works was, however, so vague that many different plants could be identified with the descriptions, and botanists attempted to press not only European, but also American, plants through the classical form. It is also important to realize that many plant genera are common to both sides of the Atlantic, and many of these are visually distinctive and easily recognized species. When Cornut discusses the use of wild ginger (*Asaron canadense*), it is more than reasonable that he refers to the use of plants of the same genus by Galen and Dioscorides. The first plants that the explorers noted, however, were those that were not like the familiar plants at home, and it was these strange new forms, not just of different colours or slightly different shapes, but totally unknown, that made the botanists question their Dioscorides and see the New World vegetation as new, and possibly as a novel source of cures for the ailments not only of Native people but of Europeans.

Unlike most animals, many plants could be removed from their native habitats and grown in Europe. Columbus had marvelled at the variety of plant life 'so different from ours' in the West Indian islands on which he made his first landfall, and a generation later maize was being grown throughout southern Europe. The tomato arrived in 1523 from Mexico, but was not widely used as a foodstuff except in the Mediterranean area.[23] Within another generation, these plants were joined by capiscum peppers, a variety of squash, tobacco, and the 'African' marigold. Not only were these exotic plants propagated, they were illustrated. An ear of corn had been depicted by Oviedo y Valdés in 1535, and then again, along with a prickly pear and another cactus, in Ramusio's *Navigationi et Viaggi* of 1606. Fuchs included maize and a pumpkin in *De Historia Stirpium* in 1542, but seemed unaware of their New World origin, calling them 'Turkish wheat' and 'Turkish cucumber.'[24] Dodoens pictured a tobacco plant in a herbal of 1554, and we have already noted Gesner's drawing of tobacco, dated *ca* 1544–5. The interest of the herbalists was not, however, only in the collection of new varieties and exotics, but also in their role as simples, or medicines.

Cartier would seem to have had a countryman's interest in plants. Much of the vegetation he saw was reminiscent of that of France, and he enumerated the various edible berries and useful trees his expedition discovered. On his 1538 voyage he included 'two Apothecaries, each with an assistant, to identify plants and determine their uses.'[25] Thevet records their delight in the novel vegetation: 'I know that certain herborists, who sought relaxation by going some way into the interior of the isle to ferret out its singularities, came back loaded with a number of exquisite plants which they made much over, regretting that conditions did not permit their immediately making a second foray.'[26] Cartier had reason to include the apothecaries. On his previous voyage, his crew had succumbed to scurvy, saved only by the administration of a decoction of the tree *Annedda*. The story of the cure is interesting for the light it sheds on the discovery of a new remedy:

> Thereupon Dom Agaya sent two women with our Captain to gather some of it, and they brought back nine or ten branches. They showed us how to grind the bark and the leaves and to boil the whole in water ... According to them this tree cured every kind of disease. They call it in their language *Annedda*.
>
> The Captain at once ordered a drink to be prepared for the sick men but none of them would taste it. At length one or two thought they would risk a trial. As soon as they had drunk it they felt better, which must clearly be ascribed to miraculous causes; for after drinking it two or three times they recovered health and strength and were cured of all the diseases they had ever had. And some of the sailors who had been suffering for five or six years from the French pox were by this medicine cured completely. When this became known, there was such a press for the medicine that they almost killed each other to have it first; so that in less than eight days a whole tree as large and as tall as any I ever saw was used up ...[27]

The *Annedda* has been identified with the *Thuja occidentalis L.*, or northern white cedar, also called arbor vitae, since it is a long-lived tree and its wood does not decay. Cartier brought the thuja back with him to France, where both it and the *Acer saccharum* (marsh or sugar maple, which Thevet notes was especially signed to the French, see above) could be seen in 1575 at the Royal Garden at Fontainebleau.[28]

Cartier was not, however, the only explorer to concern himself with plants. In the 1633 edition of Gerard's *Herball*, Johnson includes a description from de l'Écluse (Clusius) of Drake's root, or Contra-yerua:

> In the yeare (saith *Clusius*) 1581. the generous Knight *Sir Francis Drake* gave me at London certain roots, with three or four Peruvian Beazor stones, which in the Autumne before (having finished his voyage ...) he had brought with him, affirming them to be of high esteeme amongst the Peruvians: now for his sake that bestowed these roots upon me, I have given them the title *Drakena radix*, or *Drakes* root, and have made them to be expressed in a table, as you may here see them.[29]

Philip II of Spain sent his personal physician, Francisco Hernández, to Mexico in the 1570s to investigate the native pharmacopoeia and to make coloured drawings of plants, some of which were engraved in 1649. (The original watercolours were destroyed in a fire in 1671.) In 1577, John Frampton 'englished' the first Spanish book on American plants by Nicolas Monardes (1493–1588), calling it *Joyfull newes out of the newe founde worlde* ... The book described 'the rare and singuler vertues of diverse and sundrie Hearbes, Trees, Oyles, Plantes, and Stones, with their aplications, aswell for Phisicke as Chirurgerie, the saied beyng well applied bryngeth suche present remedie for all deseases, as maie seme altogether incredible: notwithstandyng by practize founde out, to bee true ...'[30] Frampton's book also included 'the portrature' of the plants, in woodcuts of mediocre quality, much resembling those of early herbals. The sassafras tree resembles an effort in topiary, its trunk straight, its crown an orb patterned with its very large and distinctive leaves. Sassafras was the sixteenth-century cure-all. Monardes notes (in Frampton's translation) that he 'had knowledge of this Tree, and a French manne whiche had been in those partes, shewed me a peece of it ...' 'Those partes' were Florida, and the Frenchman likely one of the survivors of the ill-fated Huguenot colony which lasted from 1564 to 1566. The Frenchman told Monardes how the Spaniards had learned from the surviving French, and 'how thei had cured them selves with the water of this merveilous Tree, and the manner which thei had in the usyng of it, shewed to them by the Indians, who used to cure theim selves therewith, when thei were sicke of any grief ... and it did in theim greate effectes, that it is almoste incredible ...'[31] Sassafras was touted as a cure for almost anything, and its bark was noted as an especially effective treatment for venereal disease (as was Thuja). These claims sparked a sassafras craze (the price in 1602 was £336 per ton) and led to a 1603 venture to Virginia by Bristol merchants exclusively for the gathering of sassafras. Its export was also written into the charter of the Jamestown colony, but, since sassafras failed to live up to its reputation as a panacea, the trade eventually died.

With both economic and medical benefits to be considered, the inclusion in a herbal of illustrations of New World plants was obviously vital. Hieronymus Bock eventually overcame his resistance to the inclusion of images, in part in response to the need to examine 'living images' when the actual plant was unavailable. Unfortunately very few images of North American plants painted from life existed. John White painted some from life, and possibly gave personally to Gerard the drawing of a milkweed for his 1597 *Herball.* White, too, was obviously interested in plants, since Gerard also refers to him when describing the rough binde-weed, or sarsaparilla plant, as 'one Mr. White an excellent painter, who carried very many people into Virginia.'[32] It is worth looking in some detail at John Gerard's entries on the Indian swallow-woort, thorny euphorbium, and sarsaparilla, for they are revealing of the opportunities and constraints under which the sixteenth-century botanist/herbalist laboured when describing American plants.[33] Gerard identified the Indian swallow-woort as 'a kind of Asclepias, or Swallow-woort, which the Savages call Wisanck ...' Hulton points out that, since he did not mention White's comment on the drawing, 'the hearbe which the Sauages call Wysanke wherewith theie cure their wounds which they receaue by the poysoned arroes of theire enemyes ...,' Gerard must have had an uninscribed copy.[34] Be that as it may, Gerard had at least some acquaintance with the plant itself ('which is kept in some gardens by the name of Virginia Silke Grasse'), and he describes its silk, lamenting that the Native people continue to go naked though 'the earth is covered over with this silke, which dayly they tread under their feet, which were sufficient to apparell many kingdomes, if they were carefullly manured and cherished.' He includes for reference not only the cut of the White drawing, but also that of de l'Écluse, and compares the two, complaining that neither is descriptive of the living plant, and 'upon the sight of the growing and flouring plant I tooke this description ...'[35] Gerard was fortunate in being able to view the living plant, something he could not do with 'The Torch-Thistle or thorny Euphorbium.' He describes it as rising 'up to the height of a speare of twenty foot long, although the figure expresse not the same; the reason is, the plant when the figure was drawn came to our view broken ...'[36] Here the verisimilitude of the artist is constrained by the specimen available. Much of the New World's produce arrived, not for the garden, but for the apothecary shop, and Gerard, as a barber-surgeon and herbalist, had connections with a number of apothecaries from whom he received specimens. Gerard makes an interesting comment on the trade in specimens and images of exotic plants, referring, not to a New World plant, but to ginger, which, of course, was most familiar as a root. He had obviously inquired of a friend concerning the appearance of the living plant and the friend arranged for him to receive a drawing:

> How hard and uncertaine it is to describe in words the true proportion of Plants, (having no other guide than skilfull, but yet deceitfull formes of them, sent from friends, or other meanes) they best do know who have deepliest waded in this sea of Simples. About thirty years past or more, an honest and expert Apothecarie *William Dries*, to satisfie my desire, sent me from Antwerpe to London the picture of Ginger, which he held to be truly and lively drawne: I my selfe gave him credit easily, because I was not ignorant, that there had bin often Ginger roots brought greene, new, and full of juice, from the Indies to Antwerpe; and further, that the same had budded and growne in the said *Dries* Garden. But not many yeares after, I perceived that the picture which was sent me by my Friend was a counterfeit, and before that time had been drawne and set forth by an old Dutch Herbarist.[37]

Similarly, Gerard has a problem with the descriptions of 'rough Bindeweed,' a plant supposedly common to the Old World and the New. His acquaintance with rough binde-weed, or sarsaparilla, of the New World is based, not on a drawing of the whole plant, or even a dried specimen, but on its roots, the active ingredient as far as the apothecaries are concerned, and which Gerard says 'are very well knowne to all.' He complains that 'such hath beene the carelesnesse and small providence of such as have travelled into the Indies, that hitherto not any have given us instruction sufficient, either concerning the leaves, floures, or fruit ...' Monardes, he says, affirms that the sarsaparilla has long roots, which 'is as much as if a great learned man should tell the simple, that our common carrion Crow were of a blacke colour. For who is so blinde that seeth the roote it selfe, but can easily affirme the roots to be very long?' He has received contrary reports that some say it is a vine-like plant, others a small tree, and indeed of the two types of North American plants whose present-day common names are 'sarsaparilla,' one is a vine and one a small shrubby tree. He cannot in fact describe the plant with any accuracy, but at least he can comment as a herbalist on the plant's efficacy as a remedy. His comments are interesting for the manner in which the herbalist attempted to reconcile the Old World and New World plants in the pharmacopoeia:

> *Zarzaparilla* of Peru is a strange plant and is brought unto us from the Countries of the new world called America; and such things as are brought from thence, although they also seeme and are like to those that grow in Europe, notwithstanding they doe often differ in vertue and operation: for the diversitie of the soile and of the weather doth not only breed an alteration in the form, but doth most of all prevaile in making the vertues and qualities greater or lesser ... so in like manner, although *Zarzaparilla* of Peru be like to rough Binde-weed, or to Spanish *Zarzaparilla* ... it is of a great deale more force than that which groweth either in Spaine or in Africke.[38]

The Book of God's Works: The Garden in Print

Gerard's comments on the effects of soil and weather are the comments not only of a herbalist, but of a gardener, and indeed Gerard was a well-known gardener, maintaining his own garden at Holborn and supervising the gardens of William Cecil, Lord Burghley. He comments on the sweet potato '(that I bought at the Exchange in London),' that he 'had in his garden divers roots that have flourished unto the first approch of Winter, and have growne unto a great length of branches, but they brought not forth any floures at all.'[39] Gerard also comments that he grew yucca ('I had that plant' brought me that groweth in my garden, by a servant of a learned and skilfull Apothecary of Excester, named *Mr. Tho. Edwards* ...').[40] In his dedication of the *Herball* to his employer, Lord Burghley, he writes of his gardens and his book:

> To the large and singular furniture of this noble Island, I have added from forreine places all the varietie of herbes and floures that I might any way

> obtaine, I have laboured with the soile to make it fit for plants, and with the plants that they might delight in the soile, that so they might live and prosper under our clymat, as in their native and proper countrey: what my successe hath beene, and what my furniture is, I leave to the report of them that have seene your Lordships gardens, and the little plot of myne owne especiall care and husbandry. But because gardens are privat, and many times finding an ignorant or a negligent successor, come soone to ruine, there be that have sollicited me, first by my pen, and after by the Presse to make my Labours common, and to free them from the danger whereunto a garden is subject.[41]

The impetus to preserve a garden in print grew in strength throughout the seventeenth century as the study of plants moved out of the apothecary shop and into the garden. In 1629, John Parkinson published *Paradisi in Sole, Paradisus Terrestris*, the first book in English to be devoted entirely to plants valued for their beauty as opposed to their use or medicinal value, though most of his illustrations are of the traditional herbal variety. John Prest, in his history of botanic gardens, suggests that the discovery of the New World dealt a fatal blow to classical authors and turned botanists to the garden: 'as textbooks to the natural world the classics could never recover, and the result was to throw men forward into new observations in what was known as the book of God's works.'[42] The seventeenth-century botanic gardens carried a strong religious connotation. It had been assumed in the Middle Ages that the Garden of Eden had survived the flood, and early explorers still sought the First Garden. Columbus sees in the West Indies 'the earthly paradise,' and the banana is often referred to as 'Adam's Apple,' the tree of the fruit of knowledge of good and evil. The discovery of America allowed all the scattered plants of the original garden to be brought back together, so that God's works might be seen intact, at least in Europe. Gardens were divided into four quarters, each representing a continent, and here the understanding of plant geography and the influences of 'soile and weather' began to increase, as gardeners attempted to nurture the exotics which began to flood into Europe not just from the New World, but from the Middle East. The fashion for gardens affected the way in which plants were studied, the content of the descriptions, and the accuracy of the images which appeared in books. Despite his preface, Gerard is not concerned with making an original descriptive or pictorial statement about the plants of his gardens. His text is based on that of Dodoens (though much amended and enlarged in the Johnson edition), and the illustrations, as we have noted, derived from earlier works. Whereas Gerard delved into the common stock of images, other gardeners and botanists, like Cornut, required original work.

The first catalogue of the plants in the Jardin Royal, established in the late sixteenth century, was prepared in 1601 by Jean Robin (1550–1629), curator of the botanical garden of the Paris Faculty of Medicine, where Cornut trained. Botanical gardens had become attached to a number of medical schools, the earliest having been that founded at Padua in 1545, and the one at Paris first laid out by Jean Robin in 1597. The development of botanic gardens was accompanied by the establishment of the *hortus siccus*, the dried garden or herbarium. The first was thought to have been developed by Luca Ghini (*ca* 1490–1556) at Bologna, and included 300 specimens. (By 1570 Aldrovandi is reported to have collected more than 14,500 specimens, as well as 2,000 illustrations of plants.) In the herbarium the plants were preserved pressed on sheets of paper, and it is not clear when sixteenth-century naturalists speak of receiving the image of a plant whether they are referring to a drawing or to a dried plant on a sheet. Herbarium specimens, of course, while preserving the actual plant, also distorted it, twisting the shape and in most cases losing the colour. Pierandreo Mattioli (1501–1577), author of a re-edition of Dioscorides, had resorted to soaking dried specimens in warm water to revive their true colours and shapes, but the results had not always been successful. Preserving the garden in print, as Gerard suggested, gave rise to printed text catalogues, but preserving the flowers after life required a different approach. The only method which could truly render the flower's appearance – including the stages of life; details of the bud, flower, or fruit; and true colour – was a coloured drawing from nature. In this the French botanists excelled, working with skilled artists who committed their works, not to paper, but to vellum (*vélin*), a long-lasting parchment, which, when prepared correctly, offered a finish as smooth as paper and on which the opaque body colours stood out, remaining fresh and true. The collection of *vélins* began in the early seventeenth century with the works of Pierre Vallet (b. 1575), court painter. He was succeeded by Daniel Rabel (1579–1637), followed by Nicholas Robert (1614–1685). Rabel published his images in *Theatrum Florae* of 1624, a work in which a number of French poets offered what appears now to be exaggerated praise of the artist, but in the days when images were produced, not mechanically, but by human skill, the verisimilitude for which Rabel was famous was celebrated. Scudery suggested that the birds painted in miniature by this artist were so lifelike that, if the window were opened, they would fly away. François Malherbe wrote a sonnet to the flower painter, praising his work and placing him above Apelles, the great classical master. Art has triumphed over Nature, and if the poet is not mistaken, it is Flora herself who has guided Rabel's hand.[43]

The painted image became the simulacrum of the flower, and the collection of *vélins* stood for the ephemeral flowers of the royal gardens. It was not only royalty, however, who desired representations of flowers, and books destined for the large market of amateurs and enthusiasts were produced for those whose gardens were more modest. One of the earliest of these florilegia was *Hortus Floridus*, engraved by Crispijn de Passe the Younger (*ca* 1590–*ca* 1664), one of a famous family of engravers. De Passe's 'Garden of Flowers' was printed in 1614, and was available in Latin, French, Dutch, and English versions. The frontispiece to the First Book is not an elaborate allegory (as it is in Gerard)[44] but a scene of a lady in her walled garden, tending plots bearing tulips, lilies, and daffodils, and watched by a gentleman from a rose-covered balcony (plate 17). *Hortus Floridus* is a book of flowers, not for the apothecary's dusty cabinets, but for the lady's garden. The main section of the book is divided into four parts, corresponding to the four seasons, and, unlike the illustrations for the herbals, the plants are presented as if growing in the soil, attended by butterflies and other insects (plate 18). The illustrations often show plants at various stages of their growth, so that the gardener might recognize them. Thus, the illustration of crocuses shows the first tentative spear emerging, then a flower just unfolding, and finally the full-blown bloom. The plates are decorative, and the text brief and descriptive. In the English version, the title describes the 'almost incredible laboure and paine' with which the diligent author 'hath very Labouriously compiled, and most excellently performed, both in their perfect Lineaments in representing them in theire copper plates.' The *Hortus* also includes detailed instructions for 'a most exquisite manner and methode in teachinge the practisioner to painte them even to the liffe.' The copper engravings were not issued in a coloured state, so that the author felt obliged to ensure that the 'practisioner' was provided with enough information to do justice to the plates. The instruc-

PLATE 17: Crispijn de Passe (*ca* 1590–*ca* 1664), Frontispiece to the first book, 1615. Copper engraving. From Crispin de Pass, *Hortus Floridus* (Utrecht, 1615; London, 1929). The first book is described as 'Contayninge a very lively and true Description of the Flowers of the Springe.' The English version was printed by Salomon de Roy for de Pass. French and Latin editions also exist.

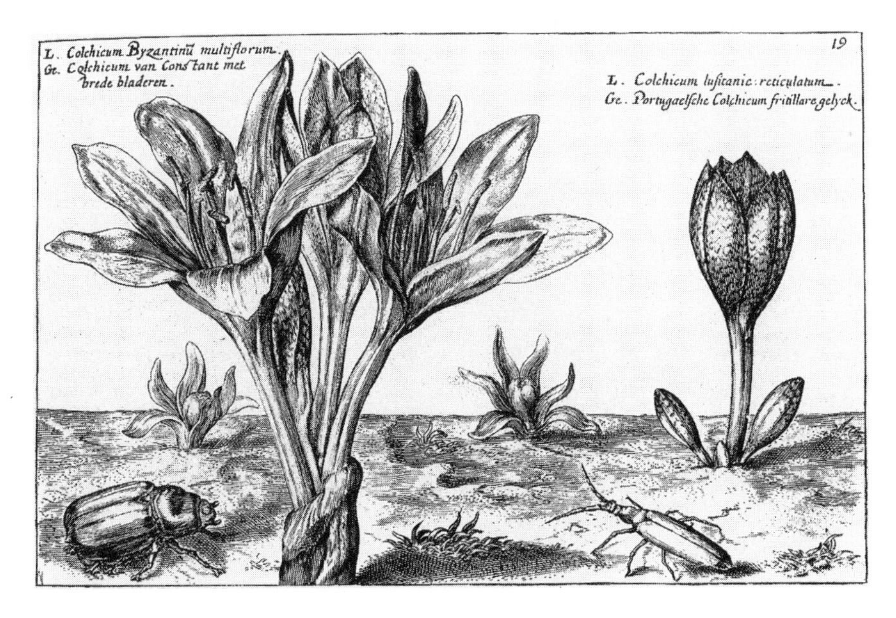

PLATE 18: Crispijn de Passe (*ca* 1590–*ca* 1664), Crocuses, 1615. Copper engraving. From Crispin de Pass, *Hortus Floridus* (Utrecht, 1615; London, 1929).

tions concerning the painting of marigolds are precise:

> The leaves that stand rounde about are of a faire masticott coloure, and if the masticot be not of a high-coloure, it must be tempered with a little lack, made shyninge, and shadovved with sad yellow, the innermost must be of a berry yellow there must be regard had in toppinge the starrs vvith the former coloure, the cowne within is the saddest of all, the leaues that come after the yellovv leaues, must very evidently appeare, because they are greene, these leaues and the steale must be shadowed vvith sad yellow and ashcoloure, and topt with white and masticott.

The author has added a verse before each season, and the verse for summer suggests that, since the engraver has used such care in describing the colours, the artist should 'In painefull paintings of the same / Good reader use no lesse.' The author also warned against 'blotts and blurrs' which would spoil the finished product, and generally harangued the reader to do justice to his original careful work.[45] The harangue was perhaps necessary, since it had not been the practice in decorative works to give flowers their proper colours. The *trompe-l'oeil* flowers that grace the borders of fifteenth-century manuscripts, while painted to look as if the living bloom had been dropped upon a page, did not necessarily sport true-to-life colours. Colour was subordinated to style and decorative appearance, and we have noted that sixteenth-century woodcuts were often crudely daubed with what can only be described as gaudy colours. The coloured frontispiece to a 1597 copy of Gerard also reveals this practice. 'In this example some flowers indeed have their rightful colours, in others truth to nature has been sacrificed to decorative ideas; all the irises of whatever species are painted in the same shades of blue, while the Madonna lily, unmistakeable from its form, is mauve blending to orange with its bulb bright red.'[46] We are again confronting the problems of resolution. The printmakers are as a rule working either from their own coloured studies or from those of others. Hans Weiditz's original colour studies were located in the herbarium of Felix Platter in Bern, Switzerland. Arber suggests that watercolour studies like these were prepared both as a guide to the printmakers and as a model for coloured copies by the publishers.[47] The black-and-white engraving or etching can only hint at the original full-colour drawing. In Dodart's *Mémoires pour servir à l'Histoire des Plantes*, published in 1675, the plates were engraved after the exquisite watercolour drawings by Nicholas Robert. Dodart notes that the engravers attempted to indicate colours through a new technique: 'Since printing in colour is not employed yet, and since painters waste much time and are not always successful, we thought we could, in future, supply to some extent what was lacking in engraving by taking care to indicate, as far as is feasible, the depth of the colour. Thus, a distinction would be made between brownish-green and pale green, between white and dark-coloured flowers ...'[48] Achieving true colour became important in the context of gardening, since gardeners often valued colour as much as shape. During the tulip craze which afflicted seventeenth-century Holland, the flower was prized for the varieties of its colour, and it became important to make these distinctions in the florilegia. De Passe remarks in relation to his figure of the 'Broad-Leaved Tulip Called Palton' that 'so great is the variety of Tulips year by year as very often to mock or surpass the desires of growers, and so it is very difficult even for one who is expert to express them in words.' Despite his protest de Passe goes on to attempt to describe the colour in words: 'this first Tulip can rightly be called *flammea*, its whole flower is resplendent, decorated with flames of sulphur-yellow ...'[49] In addition, as Arber has pointed out, the technology of copper engraving allowed the plates to be separated from the descriptions, and the plates became available for wealthy collectors and gardeners, who could look on the flower books as indoor gardens, the equivalent of the herbalist's *hortus siccus*.

Cornut and Charlevoix

While Cornut's 1632 Canadian flora is not exactly a gardening book, his illustrations for it are original, drawn from life in the gardens of Paris, and engraved, according to Blunt, in the style of Paul Reneaulme (1560–1624). Cornut's researches were not, however, limited to the Jardin Royal and the botanic garden. He also mentions the Morin gardens, which were not so much gardens as plant nurseries, and the Morins' practice of having their stock depicted in watercolours or oils was noted by John Evelyn during his 1644 visit. Cornut refers not only to the works of Dioscorides and Pliny, but also to the culture of the living flowers which he sees in these gardens. Where did the Robins and Morins acquire their Canadian plants? Cartier did indeed bring back some plants, but the gardens benefited most from the establishment in the early seventeenth century of permanent French and English settlements in North America. The 1601 catalogue cited two Canadian plants; the 1636 catalogue shows fifty North American plants. Some were undoubtedly sent by Louis Hébert (*ca* 1575–1627), an apothecary who lived first at Port-Royal, then at Quebec; also by Marc Lescarbot (*ca* 1570–1642), who had been in charge of the gardens at Port-Royal; and possibly by others, including Jesuit missionaries.[50] Others arrived via exchanges with the Tradescants of London and other gardeners. Cornut's book was not simply, then, a depiction of the plants in a particular garden,[51] but an attempt, however imperfect, to describe a regional flora. The plates do not appear to have been coloured, since they were to serve primarily for information, not for 'injoyment' as de Passe would have the flowers in his *Hortus*.

As informative images, then, Cornut's illustrations were reused over a century later in Pierre-François Xavier de Charlevoix's (1682–1761) *Histoire et description générale de la Nouvelle France*, published in 1744. De Charlevoix was not, for the most part, describing things seen in the New World for the first time; rather, in the words of M. Fournier, he uses, orders, and often

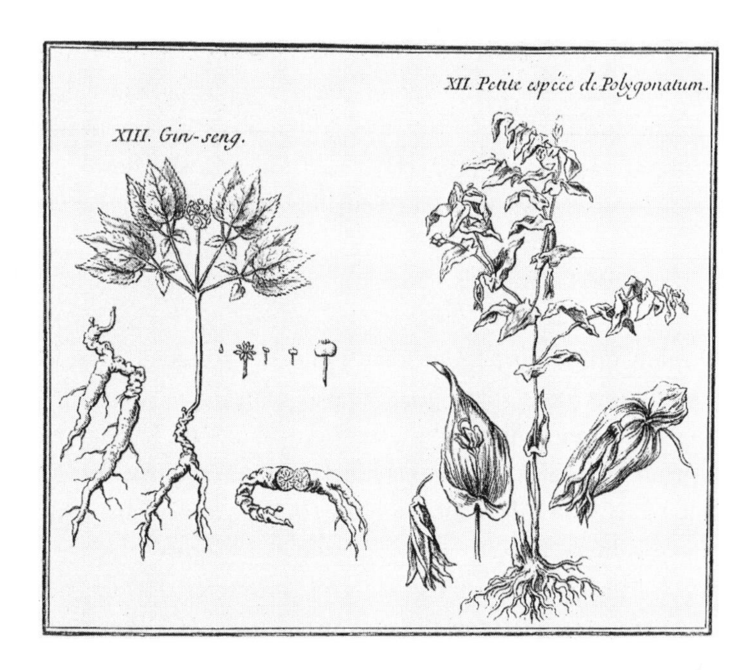

PLATE 19: New World Plants, 1744. Engraving. Four plates from Pierre-François Xavier de Charlevoix (1682–1761), *Histoire et description générale de la Nouvelle France* (Paris, 1744).

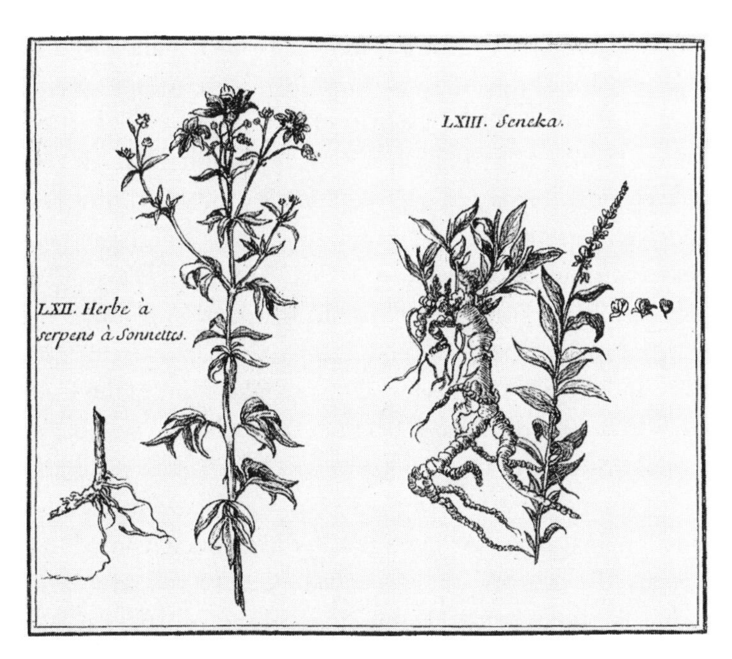
LXIII. Seneka.
LXII. Herbe à serpens à Sonnettes.

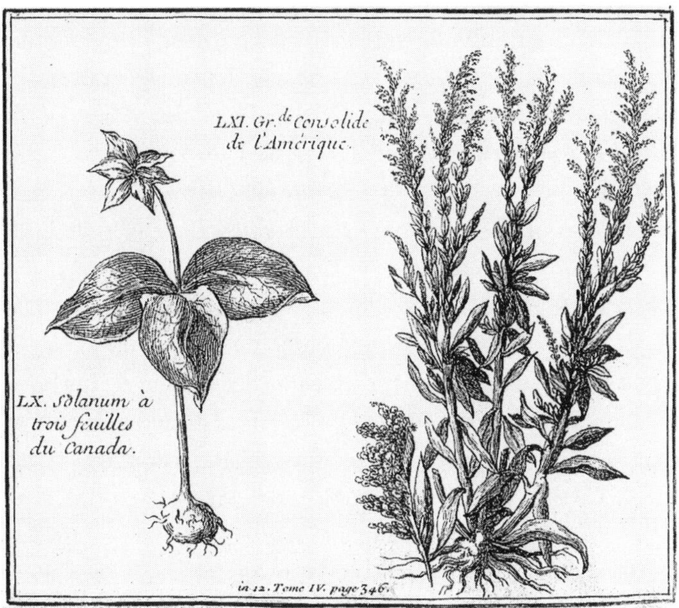
LXI. Gr.de Consolide de l'Amérique.
LX. Solanum à trois feuilles du Canada.
in 12. Tome IV. page 346.

repeats the information of earlier missionaries and other authors.[52] His descriptions of plants owe much, however, to his own observations, which were obviously recorded in a journal. In a series of 'letters' dated 1720–2 and published as volume 1 of the 1761 English edition, he describes himself as 'a traveller, rambling over the forests and plains of Canada, and who is diverted with every thing which presents itself to his view. But what could you expect from one who travels through such a country as this.'[53] In Letter IX, which is devoted primarily to natural history, he notes the distribution of trees; the manner in which Indians eat the local produce; the remedies created from bark, seed, or fruit; and the odours of plants. The second volume of the 1744 quarto edition, however, documents regional flora and includes a number of folded plates with illustrations of plants, forty of which are copied in reverse and reduced from Cornut (plate 19). In most cases, de Charlevoix has also translated Cornut's original Latin text into French, while occasionally adding his own observations. To this core, he has added an additional fifty-eight plants, incorporating information from Joseph-François Lafitau (1670–1740), Michel Sarrazin (1659–1734), Francisco Hernández, Bauhin, Tournefort, even Mark Catesby. The illustration of ginseng has been copied from Lafitau's memoir on the Canadian plant published in 1718, and Charlevoix notes that as early as 1721 Canadian ginseng was being processed in China. Many of the other plants included in de Charlevoix's North American flora are native to the southern states in which he travelled on his expedition from Canada to New Orleans. The illustrations of the additional entries differ little in style from those of Cornut. The plants show roots, some detail of flowers and fruits, and in many cases appear to be engraved after life. De Charlevoix notes, however, that, even where the notes in his own journal differ, he prefers to follow the existing illustrations.[54] It would seem that de Charlevoix worked on his book over the twenty years after his return to France, since Catesby began to publish in 1731. Most significantly, however, for our understanding of the role of images is de Charlevoix's ready acceptance of the inclusion of illustrations prepared over a century previously. The standards of accuracy in rendering which de Charlevoix, his publisher, and presumably his readers are ready to accept had not changed in a hundred years. While it cannot be denied that Cornut's illustrations are clear and often engaging, they are not engraved to the same standard as the botanical works of de Charlevoix's contemporaries. The mid-eighteenth century was the age of Georg Ehret, who has been described as the world's greatest flower painter, and his illustrations for the *Hortus Cliffortianus* of Linnaeus (1737–8) speak of a different convention and a different standard to the one used by Cornut and reused by de Charlevoix. It can only be presumed that de Charlevoix felt that, for his purposes, the illustrations of these North American plants would do very well. He was, after all, not a botanist or a taxonomist, but a curious and rambling traveller, typical of his era, and his book was not intended as a florilegium but as a description of the country and its productions. The re-engraved Cornut plates thus met the requirements, and revealed that, even in the century of Linnaeus, the difficulties inherent in producing original graphic materials could condition the use of the images and their value as information.

The slow transition from the depiction of the herbalist's dried specimens to the gardener's manuals or the glories of the florilegium reveals, as David Knight has noted concerning zoological illustration, no uniform progress towards more realistic portrayal. The plants painted by Dürer and his pupil Weiditz are extraordinarily felicitous renderings of individual plants. The needs of the apothecaries, and eventually the gardeners, could be met, however, by images which, though they may have lacked the sure touch of genius, revealed the most salient characteristics of the plant. The problems inherent in reproducing and printing images also meant that, for those who depended upon printed illustrations and who could not consult the great private collections like those of the Bishop of Eichstätt or the *vélins* of the Jardin Royal, the content of the image – its information value – triumphed over its aesthetic value. Even more significant is the acceptance by authors, publishers, and readers that an illustration once created included all the relevant information and required no amendment, even though the textual context for the image might change. The only requirement was that the image be authorized, but, as Gerard found to his dismay, authority could often be suspect. An authorized image, however, held its power for far longer than the text which had originally accompanied it. It was not until authors and artists began to see in the drawing some additional insight into the thing itself – a new language of description – that the production rather than reproduction of images achieved a new importance.

This change in the role of the image began in the seventeenth century, but it is well to remember, as Canon Raven has pointed out, that 'the popular Weltanschauung' is a collage of many parts. The demand that a plant be drawn from life and be recognizable was mediated by the repeated use of the same image no matter what the context. Like Dürer's rhinoceros, Cornut's illustrations became reference images for de Charlevoix. While these renderings were no longer emblematic, in that they stood for themselves, not for an external attribute, their repeated use decreased their information value. They stood for the plant, but could not necessarily be read as documents in themselves. Svetlana Alpers, in her analysis of the role of images in what she describes as a 'sea-change in the notion of knowledge,' refers to two Dutch prints depicting the 'marks' discovered in the core of an apple tree.[55] We have referred in both this chapter and a previous one to the idea of the mark, and of Thevet's suggestion that the maple had been 'marked' for the French by the inclusion of a fleur-de-lys in its centre. The old notion of botanical markings, or signatures, did not disappear in the seventeenth century,[56] but the idea that certain marks were also portents or omens, as Thevet suggested, did, however, come under scrutiny by the more sceptical seventeenth-century thinkers. Alpers's two prints show the tension between two ways of thinking about visual evidence. The first print shows a schematic view of an apple tree whose hollow core has taken the supposed shape of a nun or priest. Holland at this time was still at war with Catholic Spain, and the threat of invasion was present. Not long after the appearance of this first print in 1628, a second was published by Pieter Saenredam, entitled *Print to Belie Rumors about the Images Found in an Apple Tree.* This etching with text refuted the interpretation of the shapes of the hollow core as recognizable images of nuns and priests, and showed instead a dark abstract shape. The first print illustrated the mark as an omen; the second showed in detail the mark as itself. Saenredam's careful drawing was explicated in the textual description, but the image itself could be read. It was both illustration and information. Understanding, as Alpers points out, was 'located in what is seen.'[57] This notion of the visible as the ground of truth would be further explored in the eighteenth century.

four

THE REDEFINITION OF LANDSCAPE

... a vast and prodigious Cadence of Water

The curious voyage of Louis Hennepin (1626–*ca* 1705) began in 1676, when he arrived in Canada as a Récollet missionary. He spent eleven years in North America, travelling through much of la Nouvelle France and visiting Louisiana. His description of his travels in Louisiana appeared in 1683, and *Nouvelle découverte d'un tres grand pays* ... was first published in 1697. Various editions of these works comprising additional information from other published works on North America were printed in a total of forty-six editions in French, Dutch, German, Spanish, Italian, and English before the middle of the eighteenth century.[1] *Nouvelle découverte* ... had a number of illustrations, including, in the first edition of 1697, an engraved frontispiece and two plates, one of a bison and one of Niagara Falls. While the bison had often appeared before in print, this was the first published illustration of Niagara Falls for a European audience. Waterfalls and rapids exercised a particular hold on the European imagination that intensified in the era of the picturesque, of which untamed falls were an outstanding exemplar. Champlain had heard reports of the falls from the Native people, but had not himself visited them. He did, however, describe his inspection of the Saut St-Louis near Montreal, where one of his company was lost while shooting the rapids with Native companions:

> On the following day I went in another canoe to the rapid, along with this Indian, and one other of our men, to see the place where the two had

perished, and whether we could find their bodies. And I assure you that when he showed me the spot my hair stood on end to see such an awful place, and I was astonished that the victims had been so lacking in judgment as to go through such a frightful place, when they could have gone by another way ... part of this rapid was all white with foam, that made the spot so awful, with a noise so great you would have said it was thunder, and the air resounded with the sound of the falls.[2]

Champlain became accustomed to rapids, as did all voyageurs and explorers, but nothing could prepare most Europeans for the sight of Niagara. Hennepin did visit the falls with his Native guides, and his description of them is one of the highlights of his book. In the 1698 English edition, he hopes 'the Reader will be pleas'd with the Account of my Discovery; not for the Fineness of the Language, and the Nobleness of the Expression, but only upon Account of its Importance, and of the Sincerity wherewith 'tis written.'[3] Hennepin also advises his readers at the outset of the 1704 edition of his account (*Voyage curieux* ...) that the publisher had enriched this new edition with all the maps and illustrations necessary 'to give a clear Idea of certain things which are better understood, when there is some visual representation.'[4] He goes on to say that the reader would see 'a description of the Great Falls of Niagara, that are the most beautiful and most awful Cataract in the entire Universe.'[5] In case the readers would not believe his eye-witness account, he adjured, 'I protest before God that my Account is accurate and sincere, & that you can believe in all that I have reported.'[6] He described Niagara in chapter VII of the 1698 English edition of *New Discovery*:

Betwixt the Lake *Ontario* and *Erie*, there is a vast and prodigious Cadence of Water which falls down after a surprizing and astonishing manner, insomuch the the Universe does not afford its Parallel. 'Tis true, *Italy* and *Suedeland* boast of some such Things; but we may well say, they are but sorry Patterns, when compared to this of which we now speak. At the foot of this horrible Precipice, we meet with the River *Niagara*, which is not above a quarter of a League broad, but is wonderfully deep in some places. It is so rapid above this Descent, that it violently hurries down the wild Beasts while endeavouring to pass it to feed on the other side, they not being able to withstand the force of its Current, which inevitably casts them headlong above Six hundred

> foot high ... The Waters which fall from this horrible Precipice, do foam and boyl after the most hideous manner imaginable, making an outrageous Noise, more terrible than that of Thunder; for when the Wind blows out of the South, their dismal roaring may be heard more than Fifteen Leagues off.[7]

The falls were without equal in the universe; they were 'surprizing,' 'prodigious,' 'astonishing,' 'horrible.' All other falls were but feeble reflections of this wonder of the world. Hennepin obviously felt that the readers would appreciate this phenomenon only with the aid of an illustration, and the 1697 print was, for many Europeans their first 'eye-witness' vision of the landscape of Canada. The plate was, of course, executed by a European artist, likely the Dutch artist Jan van Vianen (1660–1726), but it is possible that Hennepin may have had a sketch, or even supervised the original work, since the illustration does resemble the falls (plate 20). Hennepin indeed laments (like Oviedo y Valdés before him) his lack of a trained artist to make a drawing on the spot:

> I wish'd an hundred times that somebody had been with us, who could have describ'd the Wonders of this prodigious frightful Fall, so as to give the Reader a just and natural Idea of it; such as might satisfie him, and create in him an Admiration of this Prodigy of Nature as great as it deserves. In the meantime, accept the following Draught, such as it is; in which however I have endeavour'd to give the curious Reader as just an image of it as I can.[8]

The countryside in the etching was heavily forested, as all reports agreed, with pine trees in evidence, as befit a northern country. The river that could produce so terrible a cataract was itself enormous, stretching to the far and mountainous horizon. The falls were shown as great uninterrupted cascades of water, divided by a rocky island. The 'horseshoe' shape of the Canadian falls obviously gave the artist some difficulty, or perhaps the distinctive shape was obscured by the 'third fall,' which according to a later account eventually disappeared (see below). The vantage point of the view is from the American shore, at a level with the top of the falls, but, despite Hennepin's assertion that the falls were 600 feet high, the artist could not come to terms with the scale, and from the measure of the two tiny figures perched on in the middle ground

PLATE 20: Attributed to Jan van Vianen (1660–1726), Niagara Falls, 1697. Etching. Page 44 in Louis Hennepin's *Nouvelle découverte d'un tres grand pays …* (Utrecht, 1697).

at the brink of the American falls, the cascade would appear to be about half the height of Hennepin's estimate. (The falls are in fact about 160 feet high, closer to Pehr Kalm's mid-eighteenth-century estimate of 137 French feet.)[9] The artist did, however, attempt to develop a visual metaphor for Hennepin's language in the expression of the beholders. Four figures in European dress are in the left foreground. Two are standing, gazing upon the falls; the others are seated. One of the standing figures and one of the seated figures spread their arms wide in amazement. The other two cover their ears, one with an expression of pain at the terrible noise ('more deafening than that of Thunder')[10] of the falling water. On the opposite shore, tiny figures, probably Native people, since one carries a spear, stand about near the edge of the river.

The image created by Hennepin's artist was enormously successful, since it remained in the printer's repertoire for over a century, last appearing in James Wyld's *The United States*, published in 1817.[11] Shortly after it first appeared in print, the map-maker Nicolas de Fer issued a large wall map of North and South America which included engraved vignettes by Nicolas Guérard. Hennepin's Niagara forms the background of a subsequently famous vignette in the top left-hand corner, which shows a horde of busy beavers building a dam (plate 21). The beavers, who resemble small, flat-tailed lions, and the image of Niagara are copied in reverse on the better-known map published by Herman Moll in 1715.[12] Pehr Kalm's account of the falls in *The Gentleman's Magazine* of 1751 was illustrated by a version of the Hennepin illustration, despite Kalm's disparagement of Hennepin's description, and it would appear that a hand-coloured print by Robert Hancock published *circa* 1794 combined Kalm's description with the Hennepin-based image to present an exotic Niagara to a European audience delighting in the picturesque (plate 22). Hancock's image features a group of fashionable Europeans and their dog, to whom the wonders of the falls are described by a half-naked and very dark 'savage,' clad in a loincloth and what might be an attempt at a 'feather' crown which characterized 'America' in the seventeenth century. The Europeans are accompanied by two elderly gentlemen in what appears to be Oriental dress. On the opposite shore, a line of naked 'savages' carry large burdens on their backs down a steep path. By far the most peculiar of the images featuring what became an eighteenth-century *image-clé* was, however, produced by the French engraver Sébastien Leclerc (1637–1714) in 1705 (plate 23). Leclerc produces a spectacular etching in which the truly enormous cataract becomes the

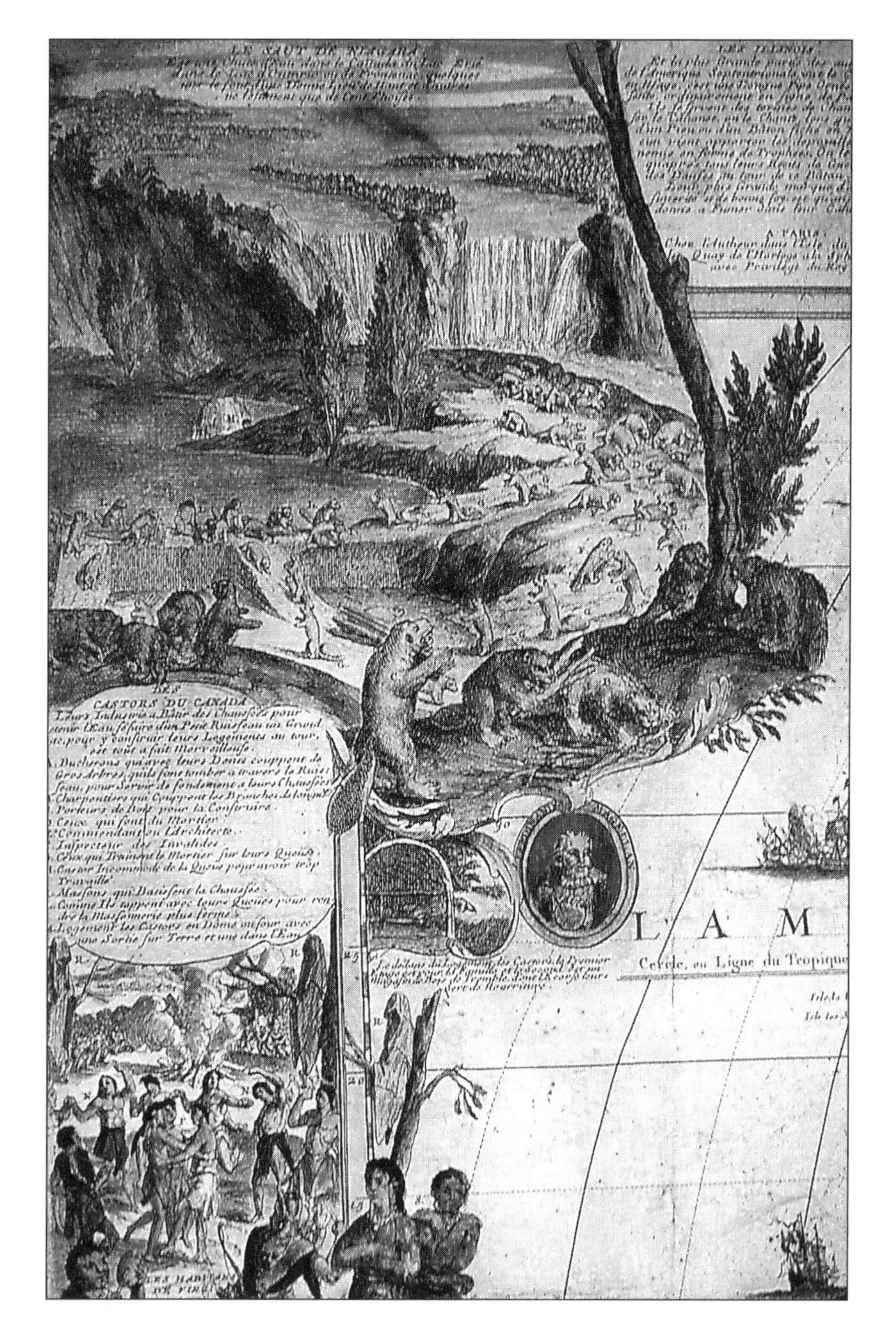

PLATE 21: Nicolas Guérard (d. 1719), Beavers and Niagara Falls, 1698. Engraving. From a vignette on Nicolas de Fer's map of North America (1698).

backdrop for the ascent into heaven of Elijah in his chariot of fire.[13] Here the Old Testament prophet has been transported from the banks of the river Jordan to the banks of the Niagara, and a New World whirlwind bends the trees and swirls the clouds, masking his ascent. The two tiny figures on the rocks below are astounded, not by the chariot of fire, but by the equally miraculous falls thundering before them. Niagara has become a visual symbol of the wonders of creation, an acknowledgment that truly all the earth is the Lord's, even so strange a landscape as that of the New World.

The Conventions of Landscape

Niagara became a symbol of the strangeness of the Canadian landscape, and a metaphor for its vast expanse. Hennepin's 1697 view was not, however, the first North American scene that Europeans had viewed, though, with the exception of some of de Bry's translations of John White's drawings, it was the most true to nature. While the animals and plants of the northern half of the New World had been figured in maps and early works, the appearance of the new land had been depicted only in the sketchiest and often most symbolic forms. Gastaldi in his early maps provided rolling hills and stylized trees. In his woodcut of a winter hunt on oddly angular snowshoes, Thevet's artist had shown only a generic snow-covered landscape with sparse deciduous trees.

Landscape per se was not, in fact, of great interest to the Europeans who first encountered the New World. Kenneth Clark makes the point, in his study of the development of landscape painting, that 'people who have given the matter no thought are apt to assume that the appreciation of natural beauty and the painting of landscape is a normal and enduring part of our spiritual activity.' He suggests, rather, that landscape painting was the chief artistic creation of the nineteenth century.[14] He does not mean by this that people did not depict the countryside before the nineteenth century, only that it was not the focus of their attention. Renaissance artists, who could paint flowers, birds, and Turkish carpets with high realism, would often show a landscape of stylized rocks to suggest mountains, or isolated trees for forests. Their interest was in the scene, not in the scenery. Similarly, early map-makers and artists illustrating texts emphasized those parts of the description their patrons and publishers felt of the greatest interest – the Native people and their costumes and customs, the animals, interesting or unusual events. The

PLATE 22: Robert Hancock (1730–1817), ***The Waterfall of Niagara***, *ca* 1794. Engraving, hand-coloured. Inscription: 'R. Hancock fecit The Waterfall of Niagara – This most surprizing Cataract of Nature is 137 feet high & its breadth about 360 yards. The Island on the middle is about 420 Yards long, & 40 Yards broad, at its lower end. The Water, on it's approaching the said island, becomes so rapid, as almost to exceed an arrow in swiftness till it comes to the Fall; where it reascends into the Air foaming white as Milk, & all in Motion like a boiling Cauldron; Its Noise may be

heard 15 Leagues off, & in Calm Weather, it's Vapours rise a great height into the Air, & may be seen like thick Smoak at 30 Miles distance. in North America.' (A French translation of the caption follows.) Hancock's description, in French and English, is based on Kalm's 1750 account. Kalm notes in this account that the falls had been measured 'with mathematical instruments' by Monsieur Morandrier, 'the king's engineer in Canada,' and that their height is precisely 137 feet.

PLATE 23: Sébastien Leclerc (1637–1714), ***Chute de la rivière de Niagara: Élie enlevé dans un char de feu, Ontario***, 1705. Etching.

landscape was sketched in like a scenic backdrop, with only the most obvious features represented. Since the landscape was not seen as information, artists could afford to use generic sketchbook trees and plants for New World species, and figure hilly terrains and high mountains, whether they existed or not. Clark notes that what he calls the 'Gothic landscape' is a landscape of symbol, not of fact, a landscape where heaps of rocks could represent mountains, and three trees make a forest. Some sense of the symbolic nature of the landscape would appear to have been retained by the map-makers and early artists, such as John White, who continued to make use of the conventions of manuscript illumination in their work.

The depiction of landscape had begun to change, however, in the late fifteenth century, and reached a new level of realism in the works of seventeenth-century Holland. The clarity of vision which permeates Dutch landscapes of the period should not, however, be confused with accurate rendering of nature. Like Dutch flower painting of the same period, many of the landscapes were studio pieces, in which the studies observed from nature were reworked into the landscapes.[15] For Clark the views of cities and countryside are portraits, the landscape of fact, the portrayal of 'recognizable experiences.'[16] When the European artists came to depict the new and largely unfamiliar landscape of the New World, they were forced to rely almost wholly on studio or printing-shop images, on the reworking of old ideas into fresh assemblages representing the new scenery. Since the scenery was still only backdrop, however, the need for factual depiction was overridden by the need for consistent style and ease of reproduction.[17] Thus the landscape of the New World was re-created in the familiar language of European illustration, and even the representation of Niagara was not so much a picture of the scenery as the portrait of a natural wonder, an emblem of unfamiliarity.

While landscape was not a focus of either textual or pictorial description, the new land did imbue all accounts with the force of its presence. The first reports described only small islands and coastlines, and in Cartier's narrative became almost an Arcadia, at least when viewed in summer. As the explorers penetrated further inland and established settlements, they began to describe the country and to depict both its strangeness and its wonders, such as Niagara or the beluga whales at the mouth of the Saguenay, as well as its familiar aspects. It should be noted once again that the northern half of the New World did not exhibit for Europeans the same wrenching exoticism as did tropical America. Many of the plants and animals were identical or very similar to those in northern Europe. The European response to the landscape, then, was not to see it as bizarre, but rather as not similar but at least recognizable. The artists used European conventions to depict the landscape not only because they had no other first-hand studies, but also because those conventions fit with the landscape that the writers described.

The Deer Park

As in most other areas, Theodor de Bry and his sons have provided the most informative early illustrations of landscape, and two plates from the later volumes in the *America* series permit insight into the seventeenth-century understanding of the *paysage* of the New World. Volume 10 of *Les grands voyages*, published in 1618 with illustrations by Johann Theodor de Bry (1561–1623), includes John Smith's accounts of Virginia. Plate XI is a hunting scene, showing the English explorers hawking, shooting, deer hunt-

PLATE 24: Johann Theodor de Bry (1561–1623), Hunting scene, 1618. Copper engraving. Plate XI from volume 10 in the *America* series or *Les grands voyages*, German edition, published by Johann Theodor de Bry (Oppenheim, 1618).

ing, and fishing. In the left foreground, a hound laps at a freshwater spring, and his master sits astride a well-groomed horse, his hawk on his wrist (plate 24). A huntsman on foot with a long gun and a sword and carrying a brace of birds walks towards his fellow. Another huntsman with his hound is releasing his hawk from the jesses, while another hawk attacks a heron, bringing it to the ground. In the mid-ground two more huntsmen with hawks are seen, one striding elegantly along, dressed in tall hat, starched ruff, and sword, hawk on his wrist, the other gesturing to the prey his hawk has brought to earth. Near them a man fishes with a long pole, his anticipated catch not any small fry, but two monster fish, similar to those which populate the oceans on maps of the same period. In the distance, against a background of mature deciduous trees,

PLATE 25: Matthaeus Merian (1593–1650), Hunting scene, 1628. Copper engraving. Page 15 from volume 13 in the *America* series or *Les grands voyages*, German edition, published by Johann Theodor de Bry (Franckfurt, 1628).

a mounted hunter, with drawn sword and companion dog, chases a stag. The sky is alive with birds wheeling and diving. It would seem that the New World offers game aplenty for the sportsman, though, as the Plymouth Pilgrims would discover, the plenty could be illusory, and starvation rather than sustenance the rule.[18] William Wood (fl. 1629–1635), writing in the same period, notes that conditions were not so idyllic for the hunter as they might appear. In *New England's Prospect* (1634), he writes that 'there be so many old trees, rotten stumps, and Indian barns, that a dog cannot run well without being shouldershot.'[19]

The idea of a New World of plenty, however, underlies Matthaeus Merian's illustration of Native caribou hunting in de Bry's 1628 edition of Sir Richard

Whitbourne's (fl. 1579–1626) *A Discourse and discovery of the New-Found-Land* (plate 25). (The fact that two years later de Bry uses the same engraving in volume 14 of the *America* series to represent deer hunting in Mexico does not detract from its use in representing a 'typical' Newfoundland scene.) The engraving on page 15 of volume 13, published in 1628, supposedly illustrates the method by which Native people hunt the caribou. In the left foreground stands a caribou in profile, recognizable by its peculiar antlers and shaggy mane over the shoulders. It stands on a grassy slope, sprinkled with flowers and backed by conifers which resemble European cypresses more than North American firs. Laden grapevines grow along the hills. In the middle ground, two Native people lounge beside a shore fringed with reeds, while another dries the fish he has presumably caught, spreading them on the ground. In the background is the scene of the hunt. The Natives have set fire to a small tree-clad island, chasing the caribou into the water, where they pursue them in dugout canoes, shooting them with bow and arrow. Birds fly to the trees. The peaceful foreground contrasts sharply with the activity of the hunt behind it, but the image of Arcadia, where little effort yields great return, is clear. One other early image from de Bry illustrating Native methods of hunting is engraved after a drawing by le Moyne and appears in the second volume of *America*, issued in 1591. Here the Native hunters disguise themselves in deerskins, luring the stags from the forest. The stags have come to drink at a meandering stream fringed with rushes. The forest in the background is thick, but the trees are large and there is little underbrush. The deer are magnificent, high-stepping animals, and the heads of both the decoy deer and the stag lured to the water are reflected in the calm stream. Again the scene is classical in its imagery, man and deerskin blending into each other, the animals of the hunt coming innocent and trusting to their deaths. These three engravings illustrate vividly the principal elements identified by Howard Mumford Jones in fifteenth- and sixteenth-century depictions of North and Central America: 'First, the component of wonder, incarnated as it were, in the concept of islands where men do not die unless they want to, where it is always summer, where food is plentiful, and where nobody works ... These idyllic promises were crossed by tales of derring-do, and the New World image absorbed as well the enchanted fairyland of chivalric romance. Finally, ... the pictorial imagination of the Mediterranean Renaissance turned naked Indians into gods and goddesses, warriors and nymphs, and what had risen out of a dream of antiquity became a mode of picturing actuality.'[20]

Hunting was, however, not just an Arcadian pastime, but an important and sometimes central preoccupation of many Europeans. Both hawking and hunting possessed a notable literature dating to classical times and widely disseminated among the aristocracy of the period.[21] Keith Thomas, in his masterly study of attitudes to nature in England, *Man and the Natural World: Changing Attitudes in England, 1500–1800*, quotes James Cleland, an early seventeenth-century English author, that 'he cannot be a gentleman which loveth not hawking and hunting.'[22] Thomas Cockayne (1519?–1592), author of *A Short Treatise of Hunting* (1591), notes in his preface 'To the Gentlemen *Readers*' that he could say 'much more in praise of this notable exercise of hunting; by which in many other Countries men haue been and yet are often deliuered from the rauine & spoile of many wild beasts; as namely of Lyons, of Beares, of Woolues, and of other such beasts of pray ...'[23] The gentlemen adventurers and explorers took their 'notable exercise' with them to the New World. Sir Walter Raleigh described the country of Guiana with a huntsman's eye: 'There is no countrey which yieldeth more pleasure to the Inhabitants, either for their common delights of hunting, hawking, fishing, fowling, and the rest, than Guiana doth. It hath so many plaines, cleare rivers, abundance of Pheasants, Partridges, Quailes, Rayles, Cranes, Herons and all other fowle; Deare of all sorts, Porkes, Hares, Lyons, Tygers, leopards, and divers other sortes of beastes, eyther for chace, or foode.'[24] Morison notes that John Davis (*ca* 1550–1605), the discoverer of Davis Strait, took hounds for hunting on his voyage of 1587.[25] William Wood encourages prospective settlers in New England to bring their dogs: 'Yet would I not dissuade any from carrying good dogs, for in the wintertime they be very useful ...'[26] Although Wood refuses to 'speak much of the hawks' for fear of censure from experienced falconers, he does wish the New England hawks 'well mewed in England, for they make havoc of hens, partridges, heathcocks, and ducks, often hindering the fowler of his long looked-for shoot.'[27]

The same passion for the chase animated the French explorers. Champlain visits the bird island and is astounded at the number and variety of birds, but finds time to enjoy as well the 'plaisir de la chasse.'[28] Marc Lescarbot dedicates much of his long poem *A-Dieu a la Nouvelle France* (1607) to a description of the birds, beasts, and fish available to the French at Port-Royal.[29] Of all the early authors on la Nouvelle France, however, few can match the enthusiasm for the hunt shown by the Baron Louis-Armand de Lom d'Arce de Lahontan (1667–1715). His *Voyages dans l'Amerique septentrionale* proved so popular that the book was republished well into the eighteenth century in English (*New Voyages to North America*, 'done into *English*' in 1703), German, and Dutch editions, as well as reprints of the original French.[30] The *Voyages* was published, for the most part, in two volumes, and the chapters of the first volume are in the form of a series of letters to a correspondent in France, a literary conceit popular at the time. De Lahontan arrives in North America as a young nobleman, and divides his time, it would appear, between military campaigns and the hunt, chiefly in the company of Native people. He is, as the author of the preface to the 1704 French edition makes him, 'a Gentleman of curiosity and good sense ... Young and full of fire ... fatigue and danger do not dishearten him in the least ... During his travels he registered everything within the compass of a spirited Cavalier ...'[31] His Letter X describes a moose hunt, and Letter XI is '*a curious Description of the Hunting of divers Animals.*' De Lahontan writes to his European correspondent, 'I find by your Letter, that you have an agreeable relish for the curious Elk-Hunting in this Country, and that further account of our other hunting Adventures, would meet with a welcome Reception. This Curiosity, indeed, is worthy of so great a Hunts-Man as your self ...'[32] De Lahontan then goes on to describe his hunt with the Native people for ducks and geese, passenger pigeons, muskrats, groundhogs, wolverines, porcupines, otters, deer, and partridges. In his second volume he describes in more detail the animals and birds he has encountered, and in particular the beaver.

Hunting, in fact, occupied a good deal of time for early European settlers, and Marc Lescarbot notes that, under his supervision, each man worked only three hours a day in the gardens or habitation, spending the rest of the time hunting and fishing.[33] This hunting was, as Raleigh noted, as much for food as for chase. Champlain describes the site of Montreal on his third voyage in 1611 as a kind of hunters' paradise where nothing is lacking, though Champlain also lingers on the description of plants and the possibility of gardens, revealing, as always, a more than common pleasure in growing things:

> And near this Place Royale there is a small river, which leads some distance into the interior, alongside which are more than sixty arpents of land, which have been cleared and are now like meadows, where one might sow grain and do gardening ... And there are many other fine meadows which would feed as many cattle as one would wish, and there are all the varieties of wood

> which we have in our forests in France, with many vines, butternuts, plums, cherries, strawberries, and other kinds of fruits which are very good to eat ... An abundance of fish can be caught, of all the varieties we have in France, and of many other very good kinds which we do not have. Game birds too of different varieties are abundant, and animals are also numerous, such as stags, fallow deer, roebucks, caribous, rabbits, lynxes, bears, beavers, and many small animals; all of these are so abundant that during the time we were at the rapid, we lacked for none of them.[34]

Twentieth-century palates are probably unused to the wide variety of birds and beasts considered fit for human consumption in the past. John White and Thomas Harriot recorded that they not only shot and sketched the strange New World wildlife, but also ate it. Without the huge commercial herds of cattle, swine, and sheep, and the battery flocks of chickens and turkeys upon which the present-day meat industry depends, people relied for their flesh on the animals they could raise on their own holdings or those they could hunt or snare in the surrounding woods. The young man of Champlain's troop who died in the rapids had gone to hunt herons for the table. Nicolas Denys (1598–1688), in his *Description and Natural History of the Coast of North America (Acadia)*, is of all the early writers an evident gourmet. His catalogue of the birds and beasts of Atlantic Canada is as much a guide to cuisine as it is to natural history. He savours almost every bird or animal in the region, and its gastronomic possibilities are as important a characteristic as its size, colour, and habitat. He declares the common porpoise 'good to eat. Black puddings and chitterlings are made from their tripe; the pluck is excellent fried; its head is better than that of mutton, but not so good as that of veal.' The soles are 'eaten with vinegar, being themselves so fat,' though Denys also like them 'on a short boiling with good herbs and an orange.'[35] He also samples skates and dolphins, sturgeons and squids, swordfish and dogfish, as well as trout and turtles – 'Being boiled, the shell is removed: then it is skinned. It is cut into pieces and served as a stew or fricassee with a white sauce. There are no pullets which are as good as this.'[36] Denys does not restrict his diet to every imaginable fish, but also eats muskrat roasted or fried with white sauce, boiled or fried beaver, young bear ('of very excellent taste'), lynx, porcupine ('as good as suckling pig'), crows ('as good to eat as chicken'), robins ('not bad to eat'), terns and gulls. He prefers young to old herons, but draws the line at cormorants. William Wood also disliked cormorants, which he comments 'be not worth the shooting because they are the worst of fowls for meat, tasting rank and fishy.' William Wood did not, however, turn his nose up at great horned owl, which he pronounced 'being as good meat as a partridge.'[37] Gabriel Sagard (d. 1650) wrote in his *The Long Journey to the Country of the Hurons* that he also ate great auk (as did countless other sailors and explorers), which he noted was 'not inferior to any game we have.'[38] He also sampled the larks from the Isle aux Alouettes near Saguenay, which tasted, he said, the same as those in France.[39] Keith Thomas notes that Europeans were exceptionally carnivorous compared with other cultures, and, though he refers in particular to England, he notes that flesh-eating was considered part of a healthy regime. Britons were noted for their taste for meat,[40] but it is obvious from the relish with which Denys, Sagard, the Baron de Lahontan, and de Charlevoix all describe the fish, fowl, and flesh which they consume that meat-eating was also highly regarded among the French. Admittedly, the French also wax eloquent over fish as well as deer, rabbit, and bear, and de Lahontan remarks that the eggs of the American eider (*Moyaque*) 'are half as big again as a Swan's, and yet they are all Yelk, and that so thick, that they must be diluted with Water, before they can be us'd in Pancakes.'[41] De Charlevoix also notes that 'we have cranes of two colours; some quite white, and others of a light grey. They all make excellent soop.'[42]

The taste for flesh in all its forms and the relative abundance of game made the New World seem to many like a promised land, or a well-stocked deer park. Enclosed deer parks were common features of sixteenth- and seventeenth-century Europe, and managed woods and forests were more common than wildwood. John Manwood's definition of a forest in *Treatise and Discourse of the Lawes of the Forrest* (1598) is quoted by Thomas: 'a certain territory of woody grounds and fruitful pastures, privileged for wild beasts and fowls of forest, chase and warren to abide in, in the safe protection of the king, for his princely delight and pleasure.'[43] The forest was by law the property of the sovereign, but nobles with a taste for hunting could empark large areas by enclosing them with a fence or ditch to maintain a herd of deer. The descriptive texts and the rare illustrations encouraged this perception of the New World as a park, turning the densely forested North American landscape into something familiar to the European eye – the wood or enclosed chase, not, however, restricted to the aristocrat, but open to all and managed by the aristocrats of the forests, the 'savages.' De Lahontan records the Native people's view of the French, and their appreciation of their own condition:

> They ... laugh at the difference of Degrees which is observ'd with us. They brand us for Slaves, and call us miserable Souls, whose Life is not worth having, alledging, That we degrade our selves in subjecting our selves to one Man who possesses the whole Power, and is bound by no Law but his own Will ... but among them the true Qualifications of a Man are, to run well, to hunt, to bend the Bow and manage the Fuzee, to work a Cannoo, to understand War, to know Forests ...[44]

De Bry and his engravers did not have to search very far for the graphic images they used to illustrate these text descriptions of hunting in the forests of the New World. The engraved scenes in *America* are very reminiscent of the images of the hunt which decorated numerous tracts dedicated to the chase published in Europe. The 1618 engraving, in particular, resembles in the disposition of the figures and the open park-like scenery, the vignette on the title page of *Les Edicts et ordonnances des roys ...* by Saint-Yon, the Lieutenant Général des Eaux et Forêts under Henri IV, published in Paris in 1610.[45] The image of the hawk attacking a heron in this same de Bry plate is common and would appear to derive from a sketchbook or pattern-book original since it appears in an almost identical form in 1575 in *The Booke of Faulconrie or Hauking* by George Turberuile (1540?–1610?). A.M. Lysaght also notes that the watercolour of the hawk and mallard in a similar configuration is frequently found in illuminated manuscripts and embroideries.[46] While the deer in the de Bry plate are well rendered, following the traditional depictions of European harts and stags,[47] in other illustrations to exploration accounts, such as those of Champlain and de Lahontan, the deer appear as emblematic animals in schematic landscapes.

In the illustration of a Native hunt described by Champlain, deer and fox flee through an open park, chased by Native hunters beating bones together, into an arrow-shaped enclosure of neat palings (plate 26). On the other side of the enclosure wait the huntsmen with spears. Also shown are two snares: one holding a fox, the other a deer. Not surprisingly, these illustrations are very reminiscent of those in one of the classics of hunting literature, 'Livre de

PLATE 26: Huron deer hunt, 1632. Engraving. From Samuel de Champlain (1567–1635), *Les Voyages de la Nouvelle France occidentale, dicte Canada: Faits par le Sr. de Champlain* ... (Paris, 1632), 265.

Chasse' by Gaston Phébus. In this fourteenth-century manuscript, illuminations show wattle fences in a similar triangular formation leading game to the waiting huntsmen armed with spears. Another illumination shows a fox (wolf?) caught in a snare to which it has been led by the same type of wattle fencing.[48] The illustration of the deer hunt in de Lahontan's work is abstracted even further, and the fence appears as schematic line (plate 27), labelled in the English edition of 1703 'The Park.' Here the deer, each animal more or less identical – emblematic stags – enter in rows, driven by orderly ranks of Native people armed with bows. The inscription reads: 'Stags block'd up in a park, after being pursued by ye Savages.' The bison hunt, which has no European prototype, appears in another plate in de Lahontan in a strangely achronological rendering, showing a herd of docile and almost identical bison walking in rows, with vignettes of various methods of attacking the animals (plate 28). These schematic depictions are akin more to maps than to proper illustrations, and although de Lahontan did complain that, in the 1703

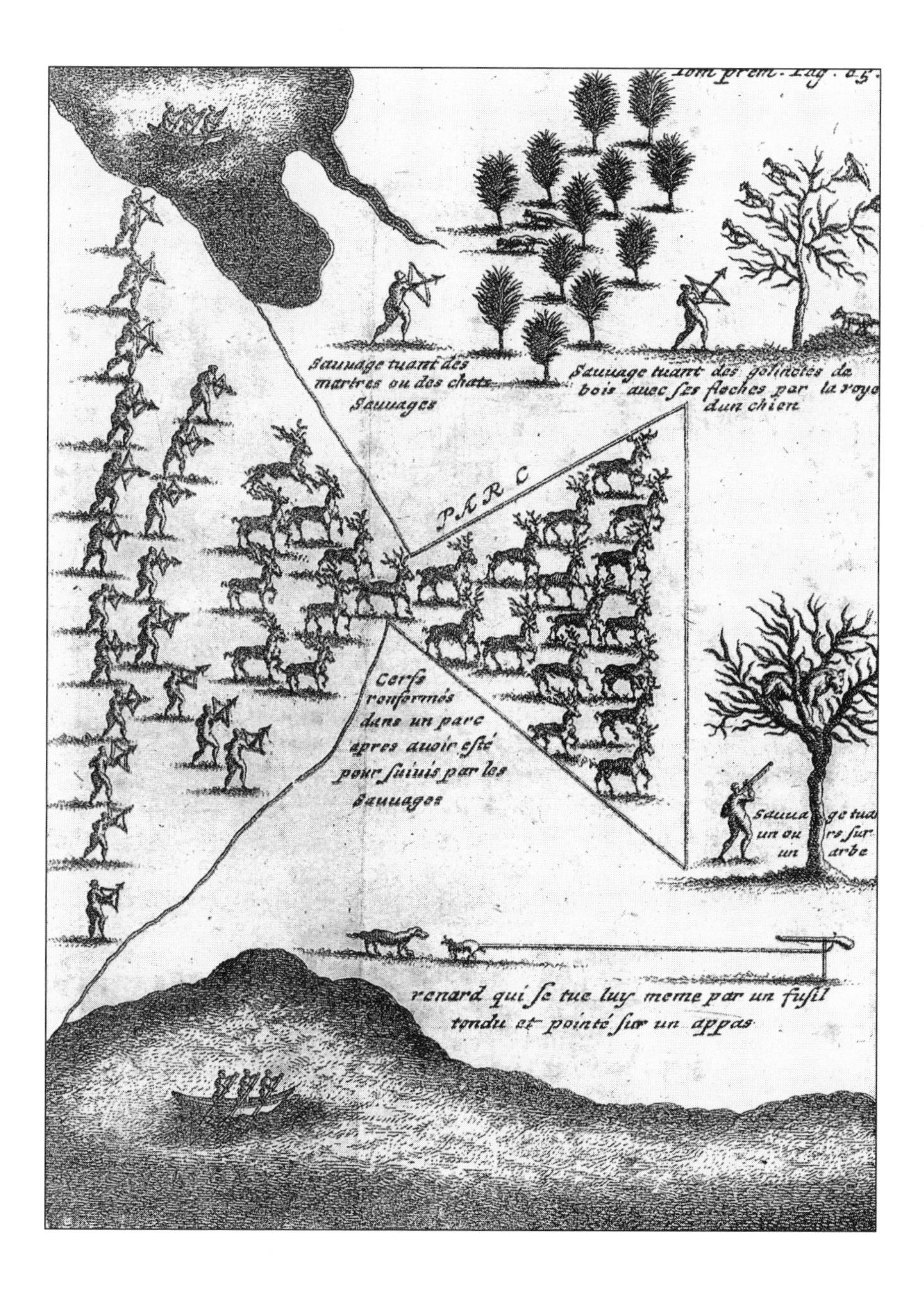

PLATE 27: Hunting, 1705. Engraving. From Baron Louis-Armand de Lom D'Arce de Lahontan, *Voyages du Baron de La Hontan dans l'Amerique septentrionale ...* (Amsterdam, 1705).

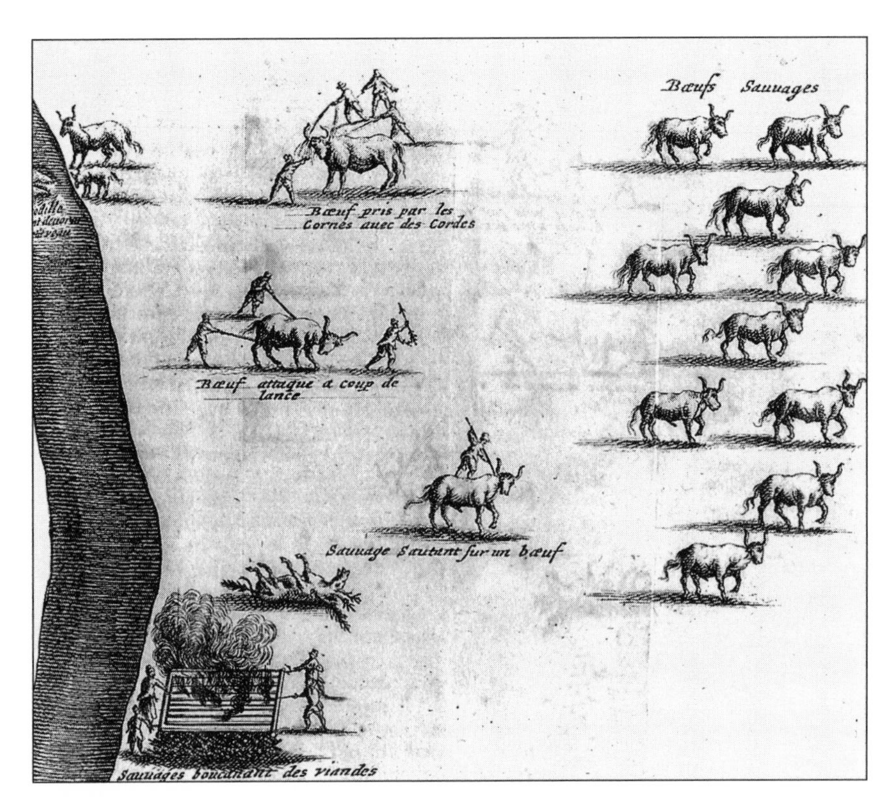

PLATE 28: Buffalo hunt, 1705. Engraving. From Baron Louis-Armand de Lom D'Arce de Lahontan, *Voyages du Baron de La Hontan dans l'Amerique septentrionale ...* (Amsterdam, 1705). Here are depicted a herd of buffalo and various Native methods of hunting and preparing the flesh. Also shown in this schematic view is a narrative incident of a 'crocodile' eating a calf.

English edition of his work, he had had to correct 'almost all the Cuts of the *Holland* Impression, for the *Dutch* Gravers had murder'd 'em, by not understanding their Explications, which were all in *French*,'[49] he did not refer to the quality of the illustrations, but to the accuracy with which they followed his text.

The Imposition of Order

Portraying the forest as a deer park or open wood was one way in which the authors and artists could comprehend the strange, heavily treed landscape of the New World. The fact that the Native people often burnt the woods to drive game or provide clearings also bolstered the illusion of a familiar managed landscape.[50] True wildwood was rarely encountered in England or France, though it of course existed in the north and the east of Europe; moreover, as a type of landscape, wilderness was not greatly valued in the seventeenth and early eighteenth centuries. As Thomas notes, when Elizabethans spoke of a wilderness they did not mean a barren wasteland, but a 'dense, uncultivated wood, like Shakespeare's Forest of Arden.' The New England colonists founded Plymouth in a 'hideous and desolate wilderness ... full of beasts and wild men ... and the whole country full of woods and thickets.' They destroyed the trees with vigour to make 'habitable' what Cotton Mather regarded as 'dismal thickets.'[51]

The strange familiarity of the northern part of the New World landscape, where Europeans could recognize trees, fruits, flowers, and animals, seems to have encouraged early explorers and settlers to attempt to see it as an Old World in the rough. They noted every meadow and clearing, every aspect that seemed to indicate that the landscape had been modified by human intervention, or that it resembled that of Europe. At Place Royale near Mount Royal in Montreal, Champlain noted that there were 'more than sixty arpents of land, which have been cleared and are now like meadows, where one might sow grain and do gardening ...,'[52] and that there are rivers 'on whose banks are many fine trees of the same varieties we have in France, with many of the finest vines I had seen anywhere.'[53] Even on his arduous journeys in 1613 through the thick forests of what is now eastern Ontario, Champlain would comment on a lake 'filled with fine large islands, which are like meadows, where it is a pleasure to hunt, venison being found there in abundance, as well as wildfowl and fish.'[54] Gabriel Sagard also praised the country of the Huron in comparison with the wilder lands he had passed through: 'It is a well-cleared country, pretty and pleasant, and crossed by streams which empty into the great lake. There is no ugly surface of great rocks and barren mountains such as one sees in many places in Canadian and Algonquin territory. The country is full of fine hills, open fields, very beautiful broad meadows ...'[55] Such pleasant open ground contrasted with the dense woods, where Champlain says they could not even portage their canoes 'on account of the thickness of the woods,'[56] and where a short passage of two-and-a-half leagues was rendered almost unbearable by the mosquitoes that molested them: 'Their pertinacity is so great that it is impossible to give any description of it.'[57] Sagard concurs with Champlain on the difficulty of passage through the woods: 'Sometimes also one has great difficulty in making a passage through dense woods, in which also a great number of trees that have rotted and fallen on one another are met with, and these one must step over. That there are rocks and stones and other obstacles which add to the toil of the trail, besides the innumerable mosquitoes which incessantly waged most cruel and vexatious war upon us ...'[58]

For some of the French brothers sent to Canada, the forests were not only thick and impenetrable, but were truly the work of the devil. Father Pierre Biard wrote that 'all this region, though capable of the same prosperity as ours, neverthless through Satan's malevolence, which reigns there, is only a horrible wilderness ...'[59] François du Creux (1596?–1666) writes, in the introduction to his history of New France, that some of his brethren even discouraged his attempts to write about Canada, which they indeed felt had no history:

> When first I undertook to compose the history of New France, many of those who have much influence with me and for whose opinion I have great respect, endeavoured to divert me from my purpose, alleging that time and labour would be lost should I select a subject at once so meagre and so grim ... it would be impossible to describe cities when cities there were none; or to mention palaces when there were no buildings but the huts of nomads, and of these but few; or to introduce well-watered gardens, pipes, canals and aqueducts when these did not exist in a land where the horror and immensity of woods and prairies occupied everything.[60]

The wildness of the New World, its disorder, and its lack of human monuments challenged the descriptive powers of seventeenth-century writers. For the most part, they simply ignored the landscape, concentrating on the narration of exploration and conquest and the customs of the Native people. When they did describe the lands and its plants and animals, they made catalogues of them, after the manner of the herbals and the pandects. The artists and engravers, even further removed from the original landscape, found it difficult to convey the chaotic nature of 'virgin forest' or even of landscape subjected to Native agricultural improvement. The depiction of homogeneous trees in neat rows, or corn planted in serried ranks in the town of Secota as engraved by de Bry in 1590 (plate 29), was not simple artistic

PLATE 29: Theodor de Bry (1528–1598), after John White (fl. 1577–1593), ***The Tovvne of Secota***, 1590. Copper engraving. Plate XX from the German edition of Thomas Harriot, *A true and briefe discourse of the new found land of Virginia*, volume 1 in the *America* series or *Les grands voyages*, published by Theodor de Bry (Franckfurt am Mayn, 1590). White's original watercolour fails to show the orderly rows of corn and other plantings. Here de Bry follows closely on Harriot's text, adding to the engraving both detail and a letter code to increase its informative value. Thus, the town is less a landscape than a mapping of both structures and activities.

convention, it also revealed a desire to bring order to the disorder of the forests that had never felt the hand of man. Thomas notes the imposition of orderly rows of trees and straight avenues over the English countryside as 'a particularly obvious way of subjecting a whole district to the authority of a great house.' The English embraced tree-planting in Elizabethan times on the grand scale, and one of the features of a great house was 'a so-called "wilderness": a dense plantation of trees, which, despite its name, was laid out in an orderly and geometrical fashion.'[61] This feeling for ordered wilderness was also expressed in the delight in an agrarian landscape with tilled fields, broad meadows, and orchards and gardens. To most Englishmen in the seventeenth century, 'a tamed inhabited and productive landscape *was* beautiful.'[62] The title page of *Hortus Floridus* (see chapter 3) is a graphic representation of the order of the enclosed Continental garden, while Francis Bacon laid down his own precepts on the ideal princely garden:

> For gardens (speaking of those which are indeed prince-like ...), the contents ought not well to be under thirty acres of ground, and to be divided into three parts: a green in the entrance; a heath or desert in the going forth; and the main garden in the midst; besides alleys on both sides ... The green hath two pleasures: the one, because nothing is more pleasant to the eye than green grass kept finely shorn; the other, because it will give you a fair alley in the midst ... The garden is best to be square, encompassed on all the four sides with a stately arched hedge ... For fountains, they are a great beauty and refreshment; but pools mar all, and make the garden unwholesome and full of flies and frogs ... For the heath, ... I wish it to be framed, as much as may be, to a natural wildness. Trees I would have none in it, but some thickets, made only of sweet-briar and honeysuckle, and some wild vine amongst; and the ground set with violets, strawberries, and primroses ... I also like little heaps, in the nature of mole-hills (such as are in wild heaths), to be set some with wild thyme; some with pinks; some with germander ... Part of which heaps to be with standards of little bushes pricked upon their top, and part without ... But these standards to be kept with cutting, that they grow not out of course.[63]

Bacon approved of a natural wildness as long as the bushes did not grow randomly.

The illustrations for *Historiae canadensis ...* of François du Creux, while not schematic, portray perhaps even more strongly than those of de Bry the sense of a managed, pastoral landscape. Du Creux was a Jesuit who prepared a ten-volume history of New France from 1625 to 1656, based on the *Jesuit Relations*. The history was published in Paris in 1664 and included in the first volume a double-page frontispiece and sixteen full-page engraved plates. Five of these plates are concerned with the natural history of Canada, and several of them convey a sense of the landscape. More so even than the Champlain or de Lahontan illustrations, the cuts are peculiarly artificial, the animals depicted in them emblematic and static. Where Champlain had described the *chaousarou*, or gar-pike, and it had been depicted with some degree of felicity on the 1613 map, du Creux accompanies his description (derived from Champlain) with an illustration of what appears to have been a mounted specimen. The fish is seen from a dorsal view with the fins correctly placed. The head appears skeletal, and it is likely that the flesh would not have been preserved on the bony head while the 'armoured' skin might survive a specimen's long journey to France. The *alce*, or elk, which is probably a moose, is unrecognizable as such, and the stiffly prancing animal resembles a generic deer, familiar in the illustrations of books on the hunt and on maps. (*Alce* was also used for the European 'elk,' but du Creux does not appear to be familiar with the Old World animal that by this time was only a memory in the less northerly parts of the continent.) He writes that 'most of the animals that are found in the forests of France are found also in Canada, but there are several that we have not yet, more especially the moose, which is by far the most famous and sometimes goes under the name of the "great beast" on account of its height.' The engraver obviously attempted to depict the beast on the basis of du Creux's comment that 'its head resembles a mule.'[64] While the *chaousarou* appears against a blank background and the mule-faced *alce* (plate 30) against a barren landscape of rock (which might be an attempt at snow, as suggested by the text), the plate of the *fiber*, or beaver, shows a New World stream lined with tall trees (plate 31). The lodge seems to be a round dome in the water in the middle ground, while the dam is an ordered barricade across the stream. The beaver themselves resemble sheared sheep, or 'mouton tondu' as Sagard (one of du Creux's important sources) describes them, with fish-scale tails, and strange three-toed bird-like feet (meant to be webbed). It is the vista of trees and rolling hills, however, that lends the engraving its European aspect.

PLATE 30: ***Alce***, 1664. Copper engraving. From François du Creux (1596?–1666), *Historiae canadensis, seu Nova-Franciae* ... (Paris, 1664), plate IV. Du Creux describes the Moose: 'There are as many beasts in the woods as there are fish in the streams. Most of the animals that are to be found in the forests of France are found also in Canada, but there are several that we have not got, more especially the moose, which is by far the most famous and sometimes goes under the name of "the great beast" on account of its height.'

PLATE 31: ***Fiber***, 1664. Copper engraving. From François du Creux (1596?–1666), *Historiae canadensis, seu Nova-Franciae ...* (Paris, 1664), plate III.

The same vista, complete with rustic cabin, appears in plate VI of the *Avis prædatrix*, or predatory bird (plate 32), in which du Creux relates that this bird, which in the engraving resembles a goose, is 'about as large as a hen, which is dun-coloured and white and black underneath. One of its feet has talons like an eagle, the other is webbed like a duck; with the latter it swims, with the former it dives into the water and brings out fish to eat, a hitherto unheard of dexterity.'[65] Du Creux is describing the osprey, a common bird along the St Lawrence, and also relatively common in Europe. His description of the bird's anatomical peculiarities, while false, was well established in the literature. This hoary legend, related in the twelfth century by Gerald of Wales,[66] and repeated some five hundred years later by Edward Topsell in *The Fowles of Heaven* (where the bird is called 'Amphibion'),[67] was no longer credible among scholars. It had been rejected definitively in 1570 by John Cay, with whom Gesner corresponded. Having chosen to render the foreign landscape conventionally and pastorally, despite the author's description of the 'rough and dense forests, which clothe the banks with their beautiful foliage,'[68] it would seem only appropriate that the illustrator add a goose-like water bird, unremarkable except for its feet.

There is only one late-seventeenth-century image that hints at the true nature of the New World forests, and that is by the same artist who appreciated the great height and breadth of Niagara. The illustrator of Hennepin's *Nouvelle découverte* ... by some means, either the accuracy of Hennepin's sketches, or perhaps even his own experience, managed to convey with some degree of accuracy the features of a genuine landscape. The engraving of the bison shows a number of interesting features (plate 33). The bison itself, while appearing too taurine, reflects accurately Hennepin's description, right down to 'an ugly Bush of Hair, which falls upon their Eyes, and makes them look horrid.'[69] The opossum hanging from a tree is recognizably an opossum, not a simivulpa, and the pelican holds its wings in characteristic fashion. It is in the background, however, that the feeling for landscape is expressed. Three bison emerge from the trees, the lead animal romping out. The forest is not a clump of stylized hardwoods, but a mixed lot of palms of two sorts, an evergreen, and deciduous trees. The woods appear thick and dark, and we can just see the last bison emerging from the cover. While the engraving is obviously a portrait of animals mentioned in the book, it has become more. It conveys the sense of the countryside of the bison's more southerly range, the scrubby subtropical forests of the southern United States.

PLATE 32: ***Avis prædatrix***, 1664. Copper engraving. From François du Creux (1596?–1666), *Historiae canadensis, seu Nova-Franciae* ... (Paris, 1664), plate VI.

PLATE 33: Attributed to Jan van Vianen (1660–1726), Buffalo, 1697. Etching. Page 187 in Louis Hennepin's *Nouvelle découverte d'un tres grand pays* ... (Utrecht, 1697).

While for most writers the North American forests were wildwood, for some they were vast woodlots awaiting harvest. Again the fact that both the eastern and the western hemispheres shared many species made a walk through the northern woods at least somewhat familiar. Cartier recognized and named a number of trees, and was amazed at single trees large enough for naval masts. De Lahontan notes that the '*Pine-trees*' are extremely tall, and ''tis said, that some of these Trees are big enough, to serve for a Mast to a First-rate Ship.'[70] The Baron describes 'The Trees & Fruits of the South Countries of Canada' and 'The Trees & Fruits of the North Countries of Canada,' commenting on their edible fruits and the quality and uses of their timber. He praises the charming aspect of Lake Erie, 'for assuredly 'tis the finest Lake upon Earth,' its banks 'deck'd with Oak-Trees, Elms, Chestnut-Trees, Walnut-Trees, Apple-Trees, Plum-Trees, and Vines which bear their fine clusters up to the very top of the Trees, upon a sort of ground that lies as smooth as one's

Hand.'[71] De Charlevoix's descriptions of the individual species are much the same and also include horticultural and pharmacological information. De Charlevoix, however, also comments on the forests as a whole, and his description, published in the mid-eighteenth century, reveals the beginnings of a change in attitude to the wildwood and its riches that will be examined more closely in a subsequent chapter:

> I don't know, Madam, whither I ought to entertain you with an account of the forests of Canada. We are here surrounded with the vastest woods in the whole world; in all appearance, they are as ancient as the world itself, and were never planted by the hand of man. Nothing can present a nobler or more magnificent prospect to the eyes, the trees hide their tops in the clouds, and the variety of different species of them is so prodigious, that even amongst all those who have most applied themselves to the knowledge of them, there is not perhaps one who is not ignorant of at least half of them. As to their quality and their uses to which they may be applied, their sentiments are so different, both in the country in which we now are, as well as in that where your grace is, that I despair of ever being ever able to give you the information I could desire on this head.[72]

For de Charlevoix, these New World forests, unlike the managed woodlots of the Old World, were never planted by the hand of man. They are part of the original creation, noble, magnificent, not habitat of Satan, but a treasure trove which Europeans in the eighteenth century might begin to put to good use.

Meta Incognita

There was one landscape, however, to which most Europeans would never become accustomed. While much of North America in the middle latitudes was reminiscent of Old World habitat, the arctic regions presented as strange a picture to most European explorers as did the tropics. Dionyse Settle authored *A true reporte of the laste voyage into the West and Northwest regions ..., worthily atchieved by Capteine Frobisher* ... in 1577 and warned his readers that 'who so maketh Navigations to these contries, hath not only extreme winds, and furious Seas, to encounter withall, but also many monstrous and great Islands of yce: a thing both rare, wonderfull, and greatly to be regarded ...'

PLATE 34: ***Inhabitants of North America near Hudsons Bay with their manner of Killing Wild Fowl***, 1768. Engraving.

When they arrived, they would find 'very little plaine ground, and no grasse, except a little ... There is no wood at all. To be briefe, there is nothing fitte, or profitable for the use of man, which that Countrie with roote yieldeth, or bringeth forth ...' Settle noted that they did find spiders '(which, as many affirme, are signes of great store of Golde).'[73] A number of editions of Settle's book were illustrated with a woodcut probably after a drawing by John White or the Flemish painter Lukas de Heere (1534–1584). White had painted watercolours of the Inuit who had been brought back to England by Frobisher, and either he or de Heere recorded their manner of hunting with a spear-thrower in a kayak in Bristol Harbour. The 1577 woodcut shows an imaginary arctic landscape of piled rocks, the sea full of duck-like birds. The Inuit costume, the kayak, and even the use of sled dogs are rendered with reasonable accuracy, but undergo remarkable modifications in subsequent copies. The 1580 cut, which appears in both a Latin edition printed in England and a German version printed at Nürnbeg (Nuremberg), shows significant changes in the disposition of figures and birds, and in the softening of the landscape from crags to shrubbery-crowned rocks. An engraving to illustrate Drake's voyages, published in 1768, shows 'Inhabitants of North America, near Hudsons Bay, with their manner of killing Wild Fowl' (plate 34). Here the arctic landscape, which at least in the Settle illustrations was relatively rocky and barren, has been transformed into a far more lush and temperate scene. The central figure in the kayak repeats the Settle/White image, but the other inhabitants no longer wear the traditional skin clothing, but a version of ragged fur coat and cloth caps. The waterway is lined with rushes, and behind a large tent are a mass of foliage and what appears to be a palm tree, but may be an attempt at a pine. The illustrator of Jens Munk's journal of his 1619–20 voyage shows a barren landscape and what appear to be lumps of ice floating in the sea, but his Native people, rather than being clothed, are naked, as the 'savages' of the tropics had been depicted.

One of the more interesting plates which accompanied another version of the Frobisher voyages was the illustration in Thomas Ellis's *A True Report of the Third and Last Voyage into Meta Incognita* (1578). The *Formschneider* here depicted an iceberg, and, though schematic in its rendering, the variable appearance suggests a cut after original sketches. The illustration comprises four views of the berg with labels, showing the iceberg's appearance as the ship first sighted it, then passed it. Perhaps it is the iceberg's strangeness, the lack of a traditional graphic vocabulary, which permitted this diagrammatic and very idiosyncratic illustration. This is not a generic but a particular object, and this explicit particularity presages the importance of accurate representation that was to become necessary in a number of the natural sciences, including geology and microscopy.

The arctic regions impressed seventeenth-century Europeans with their absolute otherness, and continued to astound most travellers in these regions until well into the nineteenth century, when their exotic scenery became a decoration on a popular series of transfer-printed British tableware. The wildwood, however, was tamed much sooner than many seventeenth-century travellers would have expected. Only the waterfalls and rapid-choked rivers remained a challenge to the European imagination and a much-pictured element of the North American landscape.

five

THE CLASSIFICATION OF THE VISIBLE: PART ONE

In 1743, George Edwards published the first of a series of illustrated books. It was entitled *A Natural History of Birds. Most of which have not been figur'd or describ'd, and others very little known from obscure or too brief Descriptions without Figures, or from Figures very ill design'd ...* The *History* was dedicated to God.[1] Part I included 'The Figures of Sixty Birds and Two Quadrupeds, engrav'd on Fify-two Copper Plates, after curious Original Drawings from Life, and exactly colour'd.' It was followed in 1747 by part II, which contained illustrations of sixty-one birds and two quadrupeds in fifty-three plates, in 1750 by part III with fifty-nine birds, and in 1751 by part IV with thirty-nine birds and sixteen animals on fifty-three plates.[2] The last two parts were published together as *Gleanings of Natural History*. All four parts were reissued in 1776 as *A Natural History of Uncommon Birds*. For *A Natural History of Birds*, Edwards, a self-taught artist and engraver, prepared an allegorical frontispiece which to contemporary eyes seems perhaps incongruous for a book of bird illustrations, but is most significant for what it reveals about Edwards and his view of the work of a natural-history illustrator (plate 35; see colour plates).

In the frontispiece, George Edwards, clothed in classical toga and sandals, sits at his easel. The easel is supported on the wings of a large bird. Before the easel sits Time, facing away from the viewer, while Minerva, goddess of wisdom, leans over Edwards's shoulder, guiding his hand on the drawing paper. Above them Juno reclines on a cloud surrounded not only by her peacock (the peacock pheasant from China), but also by a green lizard, a king vulture, an eagle, a touraco, and another exotic bird, all of whom are featured in plates in the work. A rainbow appears in the sky behind Juno, while in the background rain falls on a tall mountain (perhaps Ararat, a reminder of the flood?). At Edwards's feet are the instruments of his craft – a notebook, a sketchpad, a sheet of copper, and graving tools. Two cherubs play with brightly coloured parrots.[3] This central image is surrounded by an elaborate rococo frame ornamented with fruits and foliage as well as feathers, including the tail feathers of the peacock pheasant seen with Juno, and a bird of paradise. Several exotic birds, including a parrot and a swift, perch on the top of the frame, while the bottom bears the Latin inscription 'GEORGII EDVARDI ORNITHOLOGIA NOVA.'

Just as the emblematic frontispieces of an earlier period, like that for John Gerard's *Herbal*, were rich in hidden meanings, so does Edwards's neoclassical work repay careful analysis. Edwards's dress signals him as heir to the tradition of the great classical naturalists such as Pliny the Elder and Aristotle. At the same time, the exotic birds that perch and flutter around the figures are reminders of the impact the exploration of the New World (and exotic parts of the Old, such as Africa and China) had on classical ornithology, creating an *Ornithologia nova*, a new ornithology, and a reference to earlier work of John Ray and Francis Willughby (1676). Finally, there is the figure of Time. He looks away, for Edwards is setting down, with pencil and burin, the multitudinous and varied figures of life, preserving them in inks and washes 'for the help and information of those in future generations, that may be curious or studious in natural history.'[4] These three themes – the classical tradition, the exploration of the New World, and the importance of preservation – are important in the development of natural history and its images throughout the eighteenth century. Edwards's frontispiece is a blithe early rendering of what Bernard Smith calls the 'heroic' undertaking of the descriptive phase of the natural sciences, when Europeans set themselves the task of 'the assembling of a systematic, empirical, and faithful graphic account of all the principal kinds of rocks, plants, animals, and peoples of the world.'[5]

The Classical Tradition

While by 1700 it was obvious to most scholars that the knowledge of the classical authors did not span much beyond the Mediterranean world, many were reluctant to lose the connections with the golden age. Latin was still the language of scholarship, and many of the most important works in natural history continued to be written in the language of Pliny, chief among these the works of Carl Linné, or Linnaeus (1707–1778). The taxonomic system developed by Linnaeus relied on Latin, and sometimes Greek, words for its binomial names, but this 'system of nature' was subject to criticism by scholars concerned with the preservation of classical learning. Johann Jakob Dillenius (1687–1747), keeper of the Oxford Botanic Garden, wrote to Linnaeus in 1737 that he was displeased with the *Critica Botanica* (which Linnaeus had dedicated to him): 'We all know that botanical nomenclature is an Augean stable which ... even Gesner was unable to cleanse ... I don't object to Greek words, especially in compound names; but I think that the names of the Ancients ought not to be transferred rashly and indiscriminately to our new genera or to those of the New World. The time may perhaps come when the plants of Theophrastus and Dioscorides will be identified ...'[6] It is hard to know how many botanists shared Dillenius's faith in the omniscience of ancient authors, but for early-eighteenth-century naturalists the botanical knowledge that began with Dioscorides ran in an unbroken chain through the centuries. We have noted how Cornut and Gerard cited the classical authorities, and though the flood of exotic species would challenge the traditional schemata of botanical knowledge, it did not undermine the transmitted information that had, since Fuchs and Weiditz, been the basis of the *materia medica* and the study of plants. While new and strange specimens of animals and birds from the tropics challenged the learning of the ancients, naturalists such as John Ray could continue to mine their works because, as Ray noted in discussing the problems of migration in birds (an issue which would obsess Edwards and many of his colleagues), little advance had been made in some observations over the previous 1,600 years.[7] James Edward Smith, writing in 1791, notes the paucity of sources for the study of animals: 'It is remarkable that a part of natural history, so evidently the most important and most interesting to man, who is himself at the head of animal creation, should have lain so long uncultivated. From the time of Aristotle to Gesner and Aldrovandus, little or no improvements were made in the knowledge of animals, nor with respect to classification was any alteration attempted till the time of Ray.'[8] In the seventeenth century, the reports of careful observers in the New World were beginning to be incorporated into European works of natural history, but it was the incredible increase in the flow of specimens and materials, as well as correspondence, across the Atlantic in the eighteenth century that demanded the development of new instruments of understanding, of which the works of Edwards are prime examples.

Though clad in classical robes, Edwards's works truly comprise a 'New Ornithology' in that they break away from reliance on textual authorities and unsupported evidence. Edwards has taken Bacon to heart, and relies for his information on the thing itself, which he documents with 'full and accurate Descriptions' and 'curious Original Drawings from Life, exactly colour'd.' Edwards's descriptions, for example, have none of the fables about the anatomy of the osprey that marked Topsell's account in the *Fowles of Heaven* or are incorporated into Du Creux's *Historiae canadensis* (see chapter 4). For Edwards, first-hand accounts and accurate illustration were what distinguished his works from those of his predecessors, and, as he noted, 'everyone who consults ancient authors is very sensible of their deficiencies in the want of figures.'[9] It is the importance of illustration to Edwards and his patrons that not only distinguishes the 'new ornithology' from the old, but also suggests a new understanding of notions of representation that is prefigured in the use of manuscript drawings by the sixteenth-century pandect authors such as Gesner and Aldrovandi, whose works were still very much part of the reference base in the mid-eighteenth century. Edwards refers to Aldrovandi in particular throughout his works, and Dillenius also invoked to Linnaeus the name of Konrad Gesner. Certainly in the early eighteenth century Gesner's (in Topsell's English translation, *A History of Four-Footed Beasts ...*) and Aldrovandi's continued to be the only illustrated natural histories widely available.[10] While much of their work was based on the observations of classical authors, both Gesner and Aldrovandi developed a large correspondence and relied on colleagues for identifications, descriptions, and new specimens. Their medium of transmission, particularly for zoological specimens, was the coloured manuscript drawing, and Gesner records an instance of this practice in relation to the identification of a fish by William Turner: 'When I sent the picture of this fish to the excellent doctor William Turner of England, asking his opinion of these matters, he replied: "The illustration that you sent me is plainly of our native Lumpus, but the two rear fins are superfluous, for none such were evident in the fish I recently saw caught, and afterwards ate."'[11]

The practice of reliance on the coloured manuscript drawing for information exchange between naturalists continued uninterrupted from Gesner's day to Edwards's. George Edwards began his career as an artist for Sir Hans Sloane, whom he described as 'the good Sir Hans Sloane, Bart. who employed me for a great number of years, in drawing miniature figures of animals, &c. after nature, in watercolours, to encrease his very great collection of fine drawings by other hands ...'[12] Sloane's was not the only collection of natural-history drawings in eighteenth-century England. John Fothergill (1712–1780), Peter Collinson (1693–1768), Joseph Banks (1744–1820), Taylor White (1701–1772), and Dr Richard Mead (1673–1754), to name only a few, amassed large collections of watercolour drawings, employing, like Sloane, full-time artists to depict plants and animals. Dr John Fothergill had the largest private garden in England, and Joseph Banks noted that, in the interests of science, 'when a plant [Fothergill] had cultivated should die, he liberally paid the best artist the country afforded to draw the new ones as they came to perfection ... so numerous were they at last, that he found it necessary to employ more artists than one, in order to keep pace with their increase.'[13] In a letter to Gilbert White (1720–1793) in 1772, the Rev. William Sheffield (1742–1795) described in turn Banks's collection of drawings in natural history as 'the choicest ... that perhaps ever enriched any cabinet, public or private: – 987 plants drawn and coloured by Parkinson; and 1300 or 1400 more drawn with each of them a flower, leaf, and a portion of the stalk, coloured by the same hand; besides a number of other drawings of animals, birds, fish etc.'[14] On the Continent, the Margrave Karl III Wilhelm of Baden-Durlach collected more than 6,000 sheets of flower paintings and drawings. Georg Ehret (1708–1770) worked for a number of European collectors, in a single year (1729) producing 500 figures of plants for one ardent gardener. At the Jardin du Roy, the collection of *vélins* increased steadily as a result of the work of the Peintres au Cabinet du Roy, such as Claude Aubriet (1651–1742), the illustrator of Sebastien Vaillant's *Botanicon parisiense* (1727).

From the sixteenth century on, the coloured drawing had become accepted by naturalists as a simulacrum of the thing itself and a medium for the exchange of information. It is instructive to quote in full Edwards's comments about the importance of illustration as identifier of the thing itself:

> Natural History cannot in any degree, be perfect without figures; therefore I

> think we should promote drawing, in all such young people as seem to have a liking to it. No one need think it an amusement beneath his dignity ... Everyone who consults ancient authors is very sensible of their deficiencies in the want of the figures; for many things are mentioned by a bare name, without any description or figure; ... but there is no certain marks to shew what things in nature were called by those names, we have now wholly lost them: therefore authors, naturalists especially, should consult, first of all, the outward forms of things, in order to explain them by descriptions and other marks, and to deliver them to posterity; so as to free them, as far as human nature is capable of, from the losses and injuries they may sustain from time.[15]

Edwards's remarks return us to the subject of his frontispiece. He is not simply heir to the classical traditions of the ancient and Renaissance scholars; rather, he is the new Pliny, guided by the hand of Wisdom herself to draw and grave the perishable works of nature to preserve them against the ravages of Time's scythe. But while Pliny scorned the use of mere pictures over words, Edwards defends the role of art, which is not simply an amusement beneath the dignity of the naturalist: 'Art and nature, like two sisters, should always walk hand in hand, that so they may reciprocally aid and assist each other.'[16] Edwards's respect for the role of art in the depiction of nature is conditioned by his confrontation with the problem of the nondescript, for which no textual referents exist. Like the herbalists before him who had to classify the *herbae nudae*, Edwards must rely on the evidence of his eyes and his informants, with no previous authorities upon which to base his description. The problem of the nondescript was exacerbated during Edwards's day by the almost incredible transoceanic trade in rare specimens for the curious, and his descriptions often end with the statement that he has found no previous mention of his specimen.

The Exploration of the New World

It is almost as if, up until 1700, North America was visualized only in black and white. With the exception of some of the coloured editions of de Bry and Marcgraf and Piso,[17] the majority of seventeenth-century readers viewed the productions of the New World in uncoloured engravings. More significantly, from the point of view of natural history (other than botany), the information they gleaned from most books could only be described as incidental. Even Hennepin, whose engraved bison in a landscape provides a vivid glimpse of a new terrain, manages only a brief description of the animal's appearance, its gustatory qualities, and its migratory behaviour. As for Baron de Lahontan and other authors, the chief interest is in animals as objects of the chase and in the Native methods of hunting. In general, only the beaver is considered worthy of lengthy behavioural description – not surprisingly, given its economic importance. In 1705, everything changes with the publication in Amsterdam of Maria Sibylla Merian's *Dissertatio de Generatione et Metamorphosibus Insectorum Surinamensium.*

Maria Sibylla Merian (1647–1717) was the daughter of Matthaeus Merian the engraver, who in 1630 had provided de Bry with such a fierce New World landscape. She was trained in watercolours, however, by her stepfather, Jacob Marrel, a German painter who followed the High Renaissance style of Dürer. In 1660, Maria Merian became fascinated with insects, and notes in her journal that she collected all the caterpillars she could find to study their metamorphosis: 'I therefore withdrew from society and devoted myself to these investigations.'[18] In 1699, Merian and her youngest daughter made a two-year expedition to Surinam, where she recorded the rich insect and reptile life of the tropical forests. Her Surinam experience resulted in the sixty plates of the *Metamorphosis*, a florilegium with insects, and one of the great illustrated books of the eighteenth century. Merian was forced to publish the *Metamorphosis* with her own funds, at the urging of 'a number of amateurs of such things' who thought her insects and flowers 'the most superior and most beautiful of all the works ever painted in America.'[19] She explained her motive for publishing:

> In putting together this book, I have not sought to make money. I was simply content to cover my costs. Nor have I spared any expense to bring the work to completion. I have taken care to have the plates engraved by the most skilled artists, and to this end I have sought out the very best paper; so that I might respond not only to those who are knowledgeable about art but to all students of insects and plants. And if I find that I have achieved this goal and have satisfied and not displeased such readers, then I will indeed rejoice.[20]

Merian's efforts did not displease her designated audience, and her drawings and printed works were in the collections of Hans Sloane, Richard Mead, and James Petiver (1663–1718). The illustrations for *Metamorphosis* include spectacular full-page renditions of butterflies, moths, beetles, and flies with their larvae on the appropriate food plants. They are, as Merian says, 'all observed by me in America and drawn from life, with only a few exceptions, which I have added based on the oral testimony of the Indians.'[21] Her work was well known in the eighteenth century, and cited more than a hundred times by Linnaeus, and also used as a reference by Edwards.[22] It was a brilliant and artistic example of what might be expected in works of natural history dealing with the New World, and its imitators were not long in coming.

In 1729, Mark Catesby (1682/3–1749) began the publication of the first fully illustrated catalogue of the natural history of North America. *The Natural History of Carolina, Florida and the Bahama Islands, Containing two hundred and twenty figures of Birds, Beasts, Fishes, Serpents, Insects and Plants* was issued in two volumes (1731 and 1743) and an Appendix, its publication completed in 1747. Mark Catesby was born in Essex, but when his sister moved to America he followed, making his first trip in 1712 and returning in 1719. He collected plants for Samuel Dale (1658/9–1739), friend and colleague of John Ray's, and had a more than cursory interest in natural history. His second trip lasted from 1722 until 1725 and was undertaken with the sponsorship of a dozen patrons of natural history; he returned, finally, to London in 1726. Catesby's work owes much to the European enthusiasm for cabinets, gardening, and rare plants. James Petiver, Maria Merian's correspondent, was an avaricious and indiscriminate collector of rarities, and member of a London society (the Temple Bar Club) which sponsored plant and curio collectors in America, such as John Banister, John Lawson, and Mark Catesby. In 1700, Petiver published a broadside titled *Brief Directions for the easie making and preserving collections of all natural curiosities*, and advertised his interests to overseas travellers: 'I humbly entreat that all practitioners in Physick, Sea-Surgeons or other curious persons, who travel into foreign countries, will be pleased to make collections for me of whatever plants, shells, insects &c they shall meet with, preserving them according to directions that I have made so easie as the meanest is capable to perform, the which I am ready to give to such as shall desire them.'[23] Catesby was more than capable, and his patrons were delighted.

The extent of the American trade in ethnographic specimens, seeds, plants, animals, skins, eggs, insects, and minerals was enormous, but most of the objects found their way into private collections, to be pictured in manuscript drawings, but rarely published. Nehemiah Grew did prepare the collections of the Royal Society for publication in 1681, but the unadorned engravings of heterogeneous rarities could not command the public interest that a book like Catesby's received. Catesby's friend Cromwell Mortimer, secretary of the Royal Society, announced in the *Philosophical Transactions* that *The Natural History of Carolina* was 'the most magnificent Work I know of since the Art of Printing has been discovered.'[24] While Mortimer's praise might be seen as the admiration of a close friend, Joseph Kastner asserts that 'the *Natural History* was received as a "noble work," bought by the Queens of England and Sweden, by dozens of British peers, by Continental noblemen and, in America, by six colonial governors.'[25] The interest was in part the result of the popular enthusiasm for natural history that had been growing throughout the seventeenth century, and in part the new way in which the animals and plants were portrayed. Like Merian's, the plates were large, handsome, and gorgeously coloured. Catesby was also the first to pose his birds and beasts against botanical backgrounds, not, however, in so scientific a manner as Maria Merian. Catesby's diminutive buffalo rubs against an oversize pseudo-acacia, whose flower and leaves sprout from a single branch of a dead stump. His flamingo stands straight-necked on a beach in front of an enormous and anomalous branched gorgonian coral, whose natural height does not exceed two feet. Dr Alexander Garden (*ca* 1730–1791), a distinguished American physician and naturalist, correspondent of John Ellis (?1705–1776), was very critical of Catesby's efforts, noting that 'it is sufficiently evident that his sole object was to make showy figures of the productions of nature, rather than to give correct and accurate representations. This is rather to invent than to describe; it is indulging the fancies of his own brain, instead of contemplating and observing the beautiful works of God.'[26]

To some extent Garden's criticisms were justified. Not all Catesby's birds can be identified with current species, and certainly his contention that the birds were pictured with the plants 'of which they feed on and frequent'[27] was not always the case. In addition, there were problems with the colouring of some editions after his death. As noted above, Catesby did not render all his subjects directly after life, copying seven drawings directly from John White's originals, which were then housed in the Sloane collection. Some of his problems with 'correct and accurate representation,' however, were no doubt the result of his lack of formal training as an artist. Catesby taught himself both how to paint and, later, when he realized the cost of seeing his watercolours transformed into engravings, also how to engrave. He wrote an apology in his introduction, and it is worth quoting from it at some length, for Catesby was not the last of the self-taught artists whose works were to transform the depiction of natural history:

> As I was not bred a Painter, I hope some faults in Perspective, and other niceties, may be more readily excused: for I humbly conceive that Plants and other Things, done in a Flat, 'tho exact manner, may serve the Purpose of Natural History, better in some Measure, than in a more bold and Painter-like Way.

Similarly he excused his particular style of engraving:

> ... 'tho I may not have done in a Graver-like manner, choosing rather to omit their method of cross-Hatching, and to follow the humour of the Feathers, which is more laborious, and I hope has proved more to the purpose.

Finally, he noted some of the problems in hand-colouring:

> Of the Paints, particularly Greens, used in the illumination of figures, I had principally a regard to those most resembling Nature, ... Yet give me leave to observe there is no degree of Green, but what some Plants are possess'd of at different times of the year, and the same Plant changes its Colour gradually with it's [*sic*] Age: ... What I infer from this is, that by comparing a Painting with a living Plant, the difference of colour, if any, may proceed from the above-mentioned cause.[28]

Catesby is concerned that his readers and purchasers understand his techniques for rendering the living subject on the printed page, not only because some of his subscribers might hasten to the garden to compare their cultivated dodecatheon with the printed one, but also because he is convinced of the importance of the illustration as a means of conveying information about plants and animals. Like Merian before him and Edwards after him, Catesby acknowledges that the representation can speak more clearly than textual description: 'The Illuminating of Natural History is so particularly essential to the perfect understanding of it, that I may aver a clearer Idea may be conceived from the Figures of Animals and Plants in their proper colours, than from the most exact Description without them: wherefore I have been less prolix in the Description, judging it unnecessary to tire the reader with describing every Feather, yet, I hope I have said enough to distinguish them without confusion.'[29]

Catesby's defence of the worth of figures 'in their proper colours' calls to mind the assertions that Leonhart Fuchs made in *De Historia Stirpium* (1542). Fuchs supervised the work of the artists to ensure that the pictures were 'most perfect,' and his craftsmen 'purposely and deliberately avoided the obliteration of the natural form of the plants by shadows, and other less necessary things by which the delineators sometimes try to win artistic glory.'[30] The 'Flat, 'tho exact' and 'un-Painter-like manner' advocated by Catesby echoes the ideal of the sixteenth-century botanical draughtsmen, and like the earlier botanical compilers, Catesby was concerned with rendering unfamiliar genera to a European audience. His birds and fish and reptiles were like the *herbae nudae* of the herbalists, and he noted that 'very few of the Birds have names assigned them in the country, except some which had *Indian* names.'[31] Despite its deficiencies, Catesby's work set a new standard in illustrated natural-history books published in English, and his work remained a reference for American natural history even into the nineteenth century. Joseph Banks took Catesby with him on his trip to Newfoundland in 1762; Pennant quotes him in *Arctic Zoology* (1792); and 'Catesby's Carolina' was among the books that in 1817 Charles Fothergill (1782–1840), grandson to John, wished 'to consult when putting my Canadian researches together.'[32]

George Edwards also made extensive reference to Catesby's work, and though he specifically notes that he has 'not drawn or described any thing that was done before in any tolerable degree of perfection,' he does redraw a number of Catesby's birds, with apologies to his friend and teacher.[33] Catesby became acquainted with Edwards through the circle of curious gentlemen and women interested in natural history. Both Edwards and Catesby were self-taught and self-published, and the cost of hiring professional engravers was beyond their means. Catesby provided Edwards with specimens, and instructed him in the art of engraving or etching on copper. Like Catesby, Edwards adopted a style which he felt was best suited to the accurate portrayal of specimens which, as he reiterates, are either hitherto undescribed or have not before been figured.[34] Where Merian's elaborate drawings recall Weiditz's curled and withered leaves and the verisimilitude of Dürer, Catesby and Edwards are attempting to develop a new style best suited to the description of things. Catesby described his attempts as 'Flat, 'tho exact.' Edwards was more eloquent and also perhaps better acquainted with the prevailing aesthetic:

> Those who draw after nature, on account of natural historians, should represent things justly, and according to nature, and not strive to exalt or raise her above herself; for by so doing, instead of instructing, they will lead the world into errors; ... The historical painter, especially he that would represent the fiction of poets, may take greater liberties ...: yet every one who reads natural history, and sees figures and descriptions of things in nature, supposes they are, or ought to have been, immediately drawn and described from nature ... But in the drawing after nature, a most religious and scrupulous strictness is to be observed; and by these means we can demonstrate, that nature is or is not the same through all times. If natural historians, or they who draw for them, would carefully observe these rules, some of them might produce figures that would be deemed perfect by the knowing naturalists of these times, and escape their censure; then like the celebrated statues of Greeks and Romans, might they pass down models to future ages, as things justly and truly representing nature; but these things are rather to be wished than expected.[35]

Edwards's 'religious and scrupulous strictness' is expressed both in his desire for accurate textual description and detailed notes on sources for information and specimens, and in his rejection of some of the common stratagems employed by painters not so concerned with just representation. While he is not against posing his birds in different attitudes to add pleasing variety to his pages (in response to critics of an earlier work, possibly Albin's *A Natural History of Birds* [1731–8]), he recommends that 'many such actions, turns, and fore-shortenings, which make up the agreeable variety of masterly compositions, must be avoided, less they hide what is most conspicuous in the natural descriptions.'[36]

Throughout his works, Edwards expresses frustration with his predecessors who have not clearly depicted a bird or animal, or who made mistakes in their rendering. Eleazar Albin (fl. 1713–1759) in particular is singled out for criticism, but Catesby and Merian are also corrected, though more gently. Edwards wants to render the animal so exactly that, like the perfection of classical sculptures, his images will stand as the models for future generations. He wishes to be not only the new Pliny, but also the new Apelles, whose renditions were so lifelike they fooled the viewer into mistaking painted images for real things. He notes that he often reworked his images, as new information and new specimens came to light. But his desires were not to be realized, since, despite his best efforts, the renditions could not capture the essence of life. In part II of the *Natural History*, he wrote: 'I formerly imagined it possible, by the highest perfection in the art of painting to deceive the eye, by performing what might be taken for nature; but, ... I plainly discover it to be impracticable.'[37] Note that Edwards says 'impracticable,' not impossible. Edwards's work in depicting the nondescript is hampered both by the materials from which he must work and by the problems of rendering them on the printed page. In the next section, we examine again the problems of resolution which had plagued natural-history illustrators since the sixteenth century.

The Importance of Preservation

The ideal for most natural-history illustrators was to paint from the living specimen. Both Catesby and Merian note that they have observed and painted their subjects in their native environment. In Merian's case, she made watercolour studies of individual insects and flowers on site in Surinam, many of which are preserved in the Leningrad collections acquired by Peter the Great. She also collected and pinned butterflies in small round boxes, which she packed in crates to ship home on her return. The frontispiece to the 1771 Paris edition of her works shows a charming scene of Merian seated by a window, surrounded by her specimen boxes. Assisted by cherubs, she is preparing her specimens, while in the background can be seen a woman (perhaps her daughter) pursuing butterflies with a net in a tropical landscape (plate 36). Merian completed her drawings much in the manner of the Dutch

PLATE 36: Maria Sibylla Merian (1647–1717), Frontispiece, 1771. Engraving. From *Histoire générale des Insectes de Surinam et de Toute l'Europe*, 3d ed., edited by M. Buch'oz (Paris, 1771).

flower painters, building up a composition from individual sketches. On his excursions, Catesby had 'an *Indian* to carry [his] Box' with his 'Paper and materials for Painting,' and drew 'the Animals, particularly the Birds ... while alive (except a very few).'[38] George Edwards, on the other hand, never visited the New World. He was, in fact, a closet or cabinet naturalist, the bedell to the College of Physicians, with a large and far-flung acquaintance that catered to his passion for painting natural-history subjects, particularly birds. He managed in twenty-odd years to depict some 500 animals, insects, and birds (according to his own tally), and the means by which he achieved this reveal a great deal about both the specimen trade and the problems of accurate representation that so vexed him.

An examination of Edwards's frontispiece reveals something of both his subjects and his method. Birds perch and pose on the clouds surrounding Juno, and, indeed, for a painter who rarely left the London area, Edwards had access to a surprising number of live birds as sitters. Some, like the parrot in the iron hoop depicted in the frontispiece, were exotic pets. Sailors had been bringing back popinjays from Africa and America for hundreds of years, but more recently they had also returned with even stranger birds, such as the touraco (shown next to the peacock pheasant on Juno's cloud) and the mynah. Edwards had painted the touraco truly from life at 'Colonel Lowther's house in St. James Park,'[39] while he had painted the mynah for Dr George Wharton's lady, whose pet it was.[40] With Dr and Mrs Wharton's permission, he also took a likeness for his own collection. From his account, it would appear that Edwards was well known as an animal painter and often invited to draw an unusual bird in someone's collection. He was also not shy about asking permission to paint the rarities maintained by the curious gentlemen of the great houses, or the more lowly bird fanciers. He painted the portrait of another mynah at a 'dealer in curious birds,' whom he visited regularly. The great horned owl of the Mourning Bush Tavern also sat for a study, but his chief sources for live sitters were the menageries of the gentry and aristocrats, such as the Earl of Burlington, who kept another great horned owl from Virginia alive in his park.[41] Edwards's patron, Sir Hans Sloane, also maintained a live collection, and from it Edwards painted the King of Vultures, and the Arab bustard.[42] He also relied on Dr R.M. Massey of Stepney for the living specimens of the white-tailed eagle and the spotted hawk or falcon,[43] both from Hudson Bay. He visited the Right Honourable Lady Anson, who 'obliged [him] with the sight of a cage of these birds,' the painted finches from North America.[44] While Edwards did not include a flamingo (Catesby had drawn one), this tropical native did live for some time in the kitchen of Sir Robert Walpole's house.[45] A 'Black Hawk or Falcon' which had 'pitched on a ship belonging to the Hudson's-Bay Company, in August 1739 ... and lived in London all the hard winter, 1739' also came to Edwards's attention.[46]

Birds were the most common exotic animals kept in England, in part because, as Catesby remarks, 'there is a greater variety of the feather'd kind than of any other Animals (at least to be come at).' Birds were relatively easy to observe and catch (or shoot and preserve), and they were also attractive, or, as Catesby noted, 'they excel in the Beauty of their Colours.'[47] They were not, however, the only exotic creatures to be found in and around London. Sir Hans Sloane, who must have maintained an unusual house, also played host to the quick-hatch, or wolverene, from Hudson Bay, and the monax, or marmot, from Maryland.[48] In the late 1760s, after the fall of Quebec, judging by the number of animals imported to game parks, a fashion for moose swept London society. Two females were housed in Yorkshire and Bedfordshire, respectively, two others briefly with the Duke of Richmond, while two young males were presented by General Carleton to the Duke of Richmond in 1770 and 1773, respectively.[49] Some introductions were more serendipitous. A scarlet locust was brought 'accidentally alive from the West Indies in a basket of pineapples,'[50] and its portrait secured by Edwards.

Edwards was not, however, always so fortunate as to have his subjects alive and squawking, and he was well aware of how far short his representation might fall when his only materials were a few feathers and a beak. He notes several times that he has improved a drawing originally made from a skin, when he had access to a live specimen: 'I have not always copied servilely from draughts which I designed from nature, because some of these were drawn from dead subjects, in which I could not consider the actions and gestures of them while living; yet, after having made drawings, ... I had frequently opportunity of seeing these same birds, or birds of the like genus, from which I sketched outlines, as I had opportunity, in my visits to curious gentlemen in the neighbourhood of London.'[51]

This close reading of Edwards provides a gloss on his representation of a bittern which began this enquiry. *The Bittern from Hudson's-Bay* (plate 1, see colour plates) is a misshapen sort of bird, accurately coloured but in no way alive. This in itself is surprising, given that the bittern is also a European bird. What becomes clear, however, in reading Edwards, is that he is not a field naturalist. He is interested in depiction and in description, and by the latter he means surface characteristics. Edwards rarely includes information on behaviour, food, nesting habits, or song, and indeed he could not, when his subject was newly or long dead. No one who has viewed a bittern could fail to remark on its hunched shoulders, or its deep booming voice. Edwards's interest was not in the bird itself, but in the specimen, and his plate is an accurate representation of the bird skin that James Isham sent him from Hudson Bay. It is a drawing after nature, not after life. Edwards did not, in fact, see himself as a natural-historian; rather he identified himself first and foremost as a painter of specimens. In the description of the black hawk, or falcon, he writes: 'Whether this and the foregoing be male and female, I leave to the judgment of those who understand natural history.'[52]

The demand for specimens was far greater than the demand for live animals. (In the case of live importations, one cannot help but think that live transport simply saved the cost of preparation, for sooner or later the live animal would die and become a specimen. This was particularly the case with large mammals, which were not easily preserved for shipment.) Specimens were meant for cabinets, for the study collections of the curious, to be consulted, painted, even set up and exhibited. Specimens could be prepared and shipped in a variety of different ways, depending on their nature, the availability of preservatives, and the skill and training of the field collector.[53] Collecting in the New World was rife with difficulties, as John White had found in the early days of settlement, and as the Reverend John Clayton made clear a century later in his note 'On the Beasts of Virginia' (1688):

> And indeed by Sea I lost my Books, Chymical Instruments, Glasses and Microscopes, which rendered me uncapable of making those Remarks and Observations I had designed, they were all cast away in Captain *Wins* Ship, as they were to follow me; and *Virginia* being a Country where one cannot furnish ones self again with such things, I was discouraged from making so diligent a Scruteny as otherwise I might have done, so that I took very few Minutes down in Writing.

If the loss of his instruments were not enough, Clayton also suffered from ill health, which hampered his collecting: 'I had indeed begun once whilst in that Country to have made a Collection of the Birds, but falling sick of the Griping of the Guts, some of them for want of care Corrupted, which made them fling others away that I had thoroughly aired; for I was past taking care of them my self, there remaining but small hopes of my Life.'[54] Even in the eighteenth century, Pierre Poivre (1719–1786), supplier of bird specimens to René-Antoine Réaumur (1688–1757) in Paris, complained that the colonies in which he collected were destitute of all the recommended preparations for preserving specimens:

> you will be surprised that I still follow my former practice in preserving birds, and that I have not taken up the method outlined in your last letters: but you will forgive me when you know that this method is impractical, especially in the countries where I do my collecting and where there is neither cask nor cooper, neither alum nor Salt. One has great pains to obtain even the barest essentials of life, and our colonies lack everything, no alum, no spirit-of-wine, no chemicals, etc.[55]

Poivre's peevish reply must have been a trifle galling to Réaumur, author of a pamphlet on bird preservation, which was also reprinted in English in *Philosophical Transactions* (1748). Réaumur noted that there were three principal methods for preparing bird specimens for shipment. The first involved stuffing a cleaned bird skin with either a soft or a hard mould. This required no little skill, but, when done well, was an excellent preparation. In their correspondence, the American ornithologist Alexander Wilson (1766–1813) praised the well-known collector William Bartram (1739–1823) for his mastery of the art of mounting specimens : 'Thanks for your bird, so neatly stuffed, that I was just about to skin it.'[56] Much simpler was the second method, which required gutting and washing the bird, wrapping it carefully, then placing it in a barrel or glass jar full of brandy. Kastner remarks that some specimens suffered when unscrupulous sailors drank the brandy *en route* to their destination,[57] while Joseph Banks complained that illness prevented him from examining the specimens brought to him systematically and 'has made my Bird tub a Chaos of which I Cannot Give so good an account as I could wish.'[58] After eight days, birds might be removed from the brandy, then sealed in boxes to keep out pests. Finally, field collectors could embalm their birds by emptying the skin, then filling it with aromatic spices or a drying agent such as alum or lime. Mark Catesby apparently sprinkled snuff inside his birds, then dried them in an oven and added more snuff to keep out the insects.[59] Peter Collinson (1694–1768), the London merchant and correspondent and agent of John Bartram (1699–1777), suggested that Bartram try Catesby's method: 'F[riend] John: Mr. Catesby Desposes that thou wilt look after a night Bird call'd Wipper-Will – if this can be shot and sent in its Feathers being first, bowel'd, & Dry'd in a [brick] oven, & then Tied up in Tobacco Leaves or pack'd up in Tobacco Dust ...'[60]

Edwards received his specimens in all manner of condition. He notes that the 'Smallest Green and Red Indian Perroquet' was kept in camphorated spirits in the collection of Cromwell Mortimer. Birds in spirits lost much of their colour, and Edwards observed that, when he removed this bird, which had appeared brown, and washed and dried it, the feathers showed green and orange as in his coloured etching.[61] The same occurred with other specimens in spirits. Peter Collinson wrote to John Bartram that Mark Catesby would 'thank thee very kindly for the fruit; and come they either dry, or in spirits, they will lose their colour: so pray describe it as well as thee can, that he may be qualified to paint it.'[62] Edwards took the trouble to document his methods for others: 'If any one would draw a bird preserv'd in spirits, let him take it out, wash it pretty well in warm water, and rinse it in a good quantity of cold, and let it dry gradually, and he will restore the true colour of the feathers, as far as can be; for some feathers in the glasses of spirits, I have observed to appear of colours very contrary to the true colour they are of before they were put in.'[63]

A number of birds from Hudson Bay were sent 'preserved dry' or as 'a stuffed skin well preserved' by Alexander Light, who had been sent to Hudson Bay in 1741 on account of his interest in natural history. James Isham, an employee of the Hudson's Bay Company, 'obliged [Edwards] extremely by furnishing ... more than thirty different species of birds ... the far greatest part of them being Non-Descripts.' Isham brought his birds back to London in 1745, 'stuffed and preserved very clean and perfect,'[64] and among them was the bittern. Though some of the awkwardness in its depiction might be laid to Edwards's lack of field observation, the condition of the specimen often made it difficult to achieve the correct proportions. As Edwards wrote in regard to the 'Little Black and White Duck from Newfoundland' (which he obtained from another collector), 'one cannot with certainty give the length and breadth of dried and stuffed birds when the bodies are taken out of their skins.'[65] There were other problems with prepared specimens. When he drew the toucan, he noted that, 'after this bird was dead, the colours in the bill were wholly lost and obscured, and the bare space around the eye turned black.'[66] He preferred to receive his specimens newly dead, and noted with pleasure when a gentleman sent him a bustard cock 'fresh and in fine order.'[67] Sometimes he had to make do with specimens far less perfect, and in the case of the Argus pheasant he illustrated for *Philosophical Transactions*, he noted that 'the head and legs were supplied from the curious drawing that was sent from Canton, with the bird's skin, to Dr. *Fothergill*, which had neither feet nor head adhering to it.'[68] Finally, Edwards was fortunate in receiving some specimens at all. Shipments of specimens were part of the spoils of war, and Edwards dedicated the third part of his magnum opus to Earl Ferrars, formerly Captain Shirley Washington, 'as an acknowledgement for his kind assistance in contributing a great number of birds intended for Madam Pompadour, and taken by the Captain in a *French* prize.'[69] In a tribute to the intended recipient, Edwards called a South American manakin 'The Pompadour,' and wrote that, 'it being a bird of excessive beauty, I hope the Lady will forgive me for calling it by her name.'[70]

Edwards laboured under the problems of his materials, but he also was thwarted in his desire to establish models by the problems of his method. He found it impracticable to produce illustrations of 'what might be taken for nature,' not for lack of his own skills but for lack of the tools of reproduction that would allow him to 'deceive the eye,' to reach a degree of resolution that would mirror the lifelike effects of which painting was capable. Like earlier authors, Edwards saw art as the imitation of nature, and he acknowledged the importance of an understanding of perspective: 'A Theory of this Sort is absolutely necessary in every Painter who would Imitate Nature in almost any respect.'[71] Edwards blamed the techniques of reproduction then available. He was taught the art of etching by Mark Catesby, and made a virtue from the necessity of preparing his own plates. Although he acknowledges that 'the gravings of these figures lie under some disadvantage; because, till of late years, I had no knowledge of etching or engraving ...,' he also recognizes that reliance on professional engravers created problems in true rendering:

> ... yet, by doing them myself, I have retained in the prints some perfections, which would have been wanting, had I given my original draughts to

> engravers to copy; for they often, through want of a just understanding of the meaning of those who gave them first draughts, go a little from the author's designs, and will take some little bend and turns of strokes for the lapse of a pencil, which they will, as they purpose, correct; which sometimes robs a figure of what the author designed as its chief distinguishing mark: so that it is, in some sort, better that the original designer works such drawings on copper himself; because a man cannot so easily go from his own meaning in copying, as a second person may mistake him ...

Edwards went on to note that accuracy was particularly important in the depiction of the 'extreme parts of birds, such as the bills, and feet, and other parts' since species distinction depended on 'such little niceties.' He remarked how difficult it was for someone 'not versed in the nature of these things, to keep up a due observation and distinction of them, in copying from drawings.' Exact rendering was vital, because, even with accompanying textual description, 'it is altogther impossible for a description to give so just an idea of figures, as lines which precisely express the thing you treat of.' In his later works, Edwards had developed enough skill to draw directly on the copper plate, but he still encountered problems in the application of what Ivins has referred to as the 'web of rationality.' Not having been professionally trained, Edwards learned the art of the engraver by trial and error. He notes that his strokes in shadowing are 'not so closely and evenly laid' as those of professional engravers.[72] More significantly, he has had to develop a technique which permits the final illustration to be coloured: 'In etching plates which are afterwards to be coloured, I have discovered, that they should be done in a manner different from such things as are to continue in black and white; ... Prints that are not worked with a direct design for colouring, cannot so easily be brought to that beauty; they must be laboured and painted with body colours to make them look tolerably.'[73] Edwards is aware, in a way that clearly distinguishes him from his predecessors in the sixteenth and seventeenth centuries, of the importance of true colour in printed illustrations. Colour in natural history is, as we have observed, not trivial, and wrong colouring was as much an affront to nature as were wrong lines. He wrote in his last work, *Gleanings of Natural History*, that, 'in illuminated works of this kind, the value of the performance depends on the skill, diligence, and care of the author; for there is a very great difference; nay even an impossibility, precisely to express by words all the different degrees, shades, and mixtures of colours, so as exactly to convey your ideas to others: this can be done by no other way than by giving the colours themselves.'[74] Unlike Crispijn de Passe, he was not content simply to pass along instructions to his readers; rather, he attempted to ensure that all editions of his works were properly 'illuminated.' In his early work, he notes that he does not 'propose to part with any of the prints uncoloured while I live, lest they should afterwards be coloured by unskillful people, which might be a blemish to the work ...'[75] He lodged a complete coloured copy of his work at the Royal College Library to serve as a standard, and he included detailed colouring information in all his textual descriptions, in case readers might be forced to colour the plates themselves. Finally, in May 1769, he disposed of his works to James Robson, the bookseller who printed the 1776 edition, and left instructions that in order 'that my labours be handed down to posterity with integrity, truth, and exactness, I have delivered into his hands a complete set of the plates, highly coloured by myself, as a standard to those Artists who may be employed in colouring them for the Future.'[76] Edwards was then seventy-five years old.

No matter how skilful his strokes, however, Edwards was aware that he could not render in print the living creature. The shadows become too dark, and the highlights were never exact. The printing process would not allow him to achieve the degree of resolution required to imitate nature. Despite the limitations in his coloured prints that appear glaring to contemporary eyes – the stiff and unnatural poses, the awkward proportions, and the less-than-perfect colouring – in 1764 Edwards could reflect with some satisfaction that he had 'gone on pretty smoothly, without any competitors.' More significantly, the publication of his *Natural History of Birds* had been greeted with almost universal praise, without 'any considerable cavils raised against it, or criticisms made upon it.'[77] Edwards's birds were, in fact, described as one of the 'miracles of our century in the natural sciences ... nothing equal was seen in the past and will be in the future.'[78] High praise indeed, particularly when its author was Carolus Linnaeus, the 'Prince of Botanists' and the founder of modern taxonomic systems. The next chapter explores more deeply the relation among illustration, taxonomy, and the beginning of the modern understanding of the natural world, a process in which Linnaeus and the naturalist-painters of the eighteenth century played no small part.

six

THE CLASSIFICATION OF THE VISIBLE: PART TWO

Copiers of Nature

It is easy to dismiss George Edwards and his generation as mere daubers, producers of charming portraits of exotic pets and colourful curiosities. That is not, however, how Edwards and his circle saw either his work or the work of his colleagues. As Edwards noted, he had in twenty years met with little criticism or cavil against his books; rather, he had received honour and praise. Sir Hans Sloane had secured for him his position as bedell to the Royal College of Physicians. In 1750 he won the Copley Gold Medal from the Royal Society, and in the mid-1750s was elected a Fellow of both the Royal Society and the Society of Antiquaries. Both his books were published in French editions (*Gleanings* in parallel columns of French and English), 'for the use of foreigners,'[1] and had some influence on French natural-history artists such as François-Nicolas Martinet (fl. 1760–1800). Edwards's works were also reissued in German (1749–76)[2] and in Dutch (1772–81).[3] His designs were reinterpreted in the popular arts of his day, and Edwards complained that 'several of our manufacturers that imitate China ware, several print-sellers, and printers of linen and cotton cloths, have filled the shops in London with images, pictures and prints, modelled, copied, drawn, and coloured after the figures in my History of Birds, most of which are sadly represented as to shape and colouring.'[4] Imitation may be the sincerest form of flattery, and this imitative use of his original drawings of birds points both to the popularity of natural-history motifs among the public and to the widespread acceptance and familiarity of his work.

In many cases, however, the works of Edwards's fellow artists were at best pedestrian, or even misleading. The peculiar caribou of Peter Paillou (fl. 1744–1784) (plate 37) and the distorted raccoon of Charles Collins (plate 38) belie their reputations and their employment by two of the great naturalist-collectors of their day, Taylor White (1701–1772) and Thomas Pennant (1726–1798), both friends of Sir Joseph Banks's. Taylor White retained a number of artists between 1730 and 1760 to paint his collections, including both Paillou and Collins, as well as Georg Ehret, the celebrated flower painter whose work is discussed below in more detail. The paintings were drawn from life and, at White's death, numbered nearly a thousand. White employed Paillou from about 1744 to the early 1760s, when the latter went to work for Pennant on his *British Zoology*. Pennant complained that, while Paillou was 'an excellent artist,' he was too fond of 'giving gaudy colours to his subjects.'[5] Nevertheless, Pennant had Paillou paint for his hall at Downing 'several pictures of birds and animals, attended with suitable landscapes. Four were intended to represent the climates. The frigid zone and an *European* scene of a farm yard, are particularly well done ...'[6] Pennant also employed his Welsh servant Moses Griffith (1747–*ca* 1809) as an artist, though he had received, like Catesby and Edwards before him, no formal training. Despite his lack of training, Griffith's watercolours, which ornament the margins of Pennant's own copy of the *Arctic Zoology* (1792), now in the Blacker-Wood Library of Biology at McGill University, are wonderfully executed, the birds natural and the colours fresh, detailed with the skill of miniaturist, since none of the marginalia exceeds about ten centimetres in width. The work of Collins, Paillou, and Griffith, however, suffered from the same problems experienced by Edwards. They were forced to use skins, mounted specimens, or animals preserved in spirits, and the poor condition of their materials combined with their lack of formal training in anatomy and the dearth of living models often resulted in poor representations. The depiction of large mammals was especially difficult – Edwards had certainly done strange things to a caribou in part IV of *Gleanings* – and Paillou's caribou, which he probably did paint from life, betrays his inexperience.[7]

The difficulty in painting the larger mammals was undoubtedly the reason that George Stubbs (1724–1806), whom Bernard Smith calls 'that eighteenth-century master of visual empiricism,'[8] became a much-sought-after artist by those interested in exotic animals. Stubbs painted a moose for

PLATE 37: Peter Paillou (fl. 1744–1784), Male caribou, *ca* 1769. Watercolour on paper. A pair of caribou arrived in England in 1769, and Taylor White (1701–1772) arranged to have them recorded. Lysaght suggests that Collins was the artist, but Eleanor McLean of the Blacker-Wood Library at McGill University feels that the watercolours of the male and female caribou are the work of Paillou (personal communication).

PLATE 38: Charles Collins, Raccoon, *ca* 1759. Watercolour on paper.

William Hunter (1718–1783), the anatomist, and a kangaroo and dingo for Joseph Banks. Stubbs's work was grounded in the study of animal anatomy, and few would doubt his ability to render exactly any animal brought before him. His oil painting of the Duke of Richmond's moose was made in 1770 at Hunter's request, and Hunter relates that 'no pains was spared by that great Artist to exhibit an exact resemblance both of the young animal itself, and of a pair of Horns of the full grown Animal, which the General had likewise brought from Quebec and presented to the Duke.'[9] Stubbs had no living model when he painted the kangaroo, however, probably working from a mounted or, as Smith suggests, an inflated skin, but he managed to achieve a remarkable likeness to the animal and to the Australian landscape (likely based on a watercolour sketch by the artist Sydney Parkinson, who accompanied Cook and Joseph Banks to Australia in 1768). Unfortunately Stubbs was not well served by his engraver, Peter Mazell (fl. 1761–1797), who worked for Pennant, and both his moose and his kangaroo suffer from the translation to the engraving, the moose (*Arctic Zoology*, vol. 1, pl. VIII) acquiring inordinately long lashes (plate 39; see colour plates), and the kangaroo being transformed from an animal with muscular jumping feet to a misshapen creature with improbable clawed feet and unnatural haunches (*History of the Quadrupeds*, 1793).

While Edwards and the other artists who painted for the curious were esteemed by their patrons and by the public, in the words of Sir Joshua Reynolds they were not seen by the artistic community as anything other than 'copier[s] of nature.' Art was not about 'the imitation of nature,' according to Sir Joshua, and the 'mere copier of nature can never produce any thing great ...'[10] Writing in the same year that Stubbs painted the moose, Reynolds dismisses the work of natural-history artists: 'He [the student] will permit the lower painter, like the florist or collector of shells, to exhibit the minute discriminations, which distinguish one object of the same species from another; while he, like the philosopher, will consider nature in the abstract, and represent in every one of his figures the character of its species.'[11] Twelve years later, when Edwards has published *Gleanings* and Pennant has issued the first edition of *History of the Quadrupeds,* Reynolds asserts that the 'detail of particulars' is not the business of the artist but 'presupposes *nicety* and *research*, which are only the business of the curious and attentive, and therefore does not speak to the general sense of the whole species ...'[12] He goes on to lecture the students at the academy:

> A Landscape-Painter certainly ought to study anatomically (if I may use the expression) all the objects which he paints; but when he is to turn his studies to use, his skill, as a man of Genius, will be displayed in shewing the general effect, preserving the same degree of hardness and softness which the objects have in nature; for he applies himself to the imagination, not to the curiosity, and works not for the Virtuoso or the Naturalist, but for the common observer of life and nature.[13]

It was precisely 'nicety' and 'research' which were so prized by the naturalists, and by those who drew 'after Nature' or 'from Life.' We noted in chapter 5 that to draw 'after nature' did not necessarily mean to use a living model; rather, the artist who drew from nature worked according to a special set of rules by which the specimen might be transformed, for its better preservation and study, into the illustration. Edwards noted that he was at first reluctant to publish some of his drawings because he had very little information on the birds themselves, in many cases not even knowing their country of origin. His friends answered: 'that as I had taken the draughts from nature, and that it would be well attested, and the like birds might perhaps never be met with again, it was better to preserve the figures without knowing their countries than not at all.'[14] What was important, then, was not the quality of information about the bird, but the publication of the image itself. These were truly copies of nature, as his friends could attest, and not, as had been the case in the past, simply drawings from memory, from the realm of the imagination rather than from that of the trained curiosity. Hunter contrasted the work of artists such as Stubbs and Edwards with those who had worked for Aldrovandi in depicting the European elk, or *alce*:

> His figures are done with so little Art, and are so little like our Orignal, that we dare not pronounce them to have been copied from the same species. I should guess the Male figure to be ideal, and the female to have been done [by] some person who had seen the Animal, and who had made the design afterwards from memory. In the figure of the Male there is no expression of the Character and proportions of the Animal, and the mane is flowing. The female figure expresses the Character of the Animal; but the Mane is vastly too small and is continued all the way to the Tail. We cannot suppose that the painter saw the Animal at the time he made the design.

Hunter then goes on to criticize the figures of elk and 'orignal' in French publications, and adds: 'It is to be hoped that the Naturalists of the Northern Countries will give us such faithfull and expressive figures of the Elk and other animals as to leave us no longer in doubt.'[15] The eye-witness account of an animal is vital, for, as Hunter notes, the ideal image or that done from memory is inadequate for the purposes of science; moreover, the images must be executed with some 'Art,' so that they correspond visually to the living creature. Hunter also makes a curious observation on the 'Character' of the animal – namely, that the figure of the male does not exhibit the character, while that of the female exhibits the character, but is still in overall appearance incorrect. Reynolds has also made allusion to the representation in the figures of the landscape painter of the character of the species. What is this 'character' that both refer to, and in each case must represent the essence of that which is depicted?

'The Invention of the Present Age'

'Character' becomes a significant word in the eighteenth-century context. Appropriately for this discussion, the word itself is derived from the Greek word for a graving tool. To be stamped with a 'character' was to be marked by a brand, which gradually came to mean to exhibit a particular trait or feature. To describe the character is to describe the distinguishing mark. By the middle of the eighteenth century, its use, particularly in natural-history writing, becomes freighted with Linnaean connotations. In much of Europe, Carolus Linnaeus was the most celebrated naturalist of his day. Through his own writings and the work of his many pupils and followers, the system of classification he advocated was applied, first to plants, then to all manner of living things, which, by being named, were brought within the scientific understanding of a generation. Any appreciation of the role of illustration in this period must be seen within the Linnaean context and system.

Linnaeus first published his *Systema Naturae* in 1735. In the brief two years that followed, he also published the *Fundamenta Botanica* and the *Bibliotheca Botanica* in 1735, followed in 1737 by the *Genera Plantarum*, the *Flora Lapponica*, and the *Critica Botanica*. In addition to these works, he was the author of an illustrated catalogue of George Clifford's (1685–1760) gardens at Hartekamp. The *Hortus Cliffortianus*, with drawings by Ehret and Jan Wandelaar (1690–1759), was engraved and printed at Amsterdam in late 1737. It is no wonder that the young Swedish botanist fell seriously ill in the winter of 1737–8, and was nursed back to health at the home of Clifford, for whom he had produced such a magnificent and important catalogue. The *Hortus* itself is a large folio-sized work, written in Latin, and is subtitled *Plantas exhibens quas in Hortistam Vivis quam Siccis Hartecampi in Hollandia*. The plants in Clifford's gardens were by no means the everyday plants of most Dutch gardens, and Linnaeus described in his rather florid style the impression which Hartekamp made on him: 'The fame of Your Garden, illustrious Clifford, was on the lips of a few men, but less constantly than it deserved to be, and I was persuaded that Your Garden was only a Tantalus or Hesperides, such as cover almost the whole of the intensely cultivated land of Holland; I hardly considered it worth visiting, but the actuality surpassed all expectation ... Dumbstruck I gazed. It pierced my heartstrings through.'[16]

The allegorical frontispiece by Wandelaar shows Europa seated on a lion

PLATE 40: Jan Wandelaar (1690–1759), Frontispiece, 1737–8. Engraving. From Carolus Linnaeus, *Hortus Cliffortianus* (Amsterdam, 1737–8). Jan Wandelaar worked as both artist and engraver for a number of plates in the *Hortus*. Wandelaar was also the engraver for Claude Aubriet's drawings in the *Botanicon parisiense* (1727).

and accepting the horticultural tributes of America, Africa, and Asia (plate 40). The young Linnaeus (represented as Apollo) hovers over Europe. Behind them is the entrance to the garden itself; to their left grows a banana tree, which Linnaeus managed to nurture into flower and fruit during his stay at the garden. Below Apollo's feet is the vanquished dragon or hydra, representing the faked specimen of a many-headed monster whose authenticity Linnaeus had dismissed during a visit to Hamburg. Also pictured are cupids who sport with a watering can, brazier, and garden plan, as well as with a Celsius thermometer, which Linnaeus had some claim to have invented. The *Hortus* lists the explorers ('Peregrinatores') from whom Linnaeus has sought his information and among whom he notes Monardus, Plumier, Catesby, Marcgraf and Piso, and Merian, then proceeds to the plates and their descriptions. Here the *Hortus* differs considerably from the work of Cornut, or from another describer of Parisian botany, Sébastien Vaillant (1669–1722), author of *Botanicon parisiense* (1727). Cornut's listing of plants appears to follow a random order,[17] while Vaillant lists his plants in alphabetical order. Cornut's engravings show the entire plant, with details of flower, fruit, and root, enough to be recognizable by the herbalist. Vaillant's illustrations, by Claude Aubriet, engraved by Wandelaar, are crowded together, many to a page, with details of the flowers. The illustrations prepared by Ehret and Wandelaar for the *Hortus*, on the other hand, are printed one to a page, and the number of details has increased. For example, on plate V, a collinsonia by Ehret, are included 'a. *Flos magnitudine naturali*, b. *Idem a tergo visus*, c. *Calyx sub florescentia constitutus*, d. *Idem fructu praegnans*, and e. *Germen*' (plate 41). The itura from Mexico is pictured in its pot as an '*arbor sexpedum*,' with details of its stem; the stem with leaves at the top; the flowers, though, in October 1737, '*flores tamen imperfecti decidue*'; a leaf of natural size (lightly engraved in the background); and finally a flower dissected. Some plants like the bauhinia could be shown only without flower or fruit as the plant had been just one year growing. The details added to Ehret's and Wandelaar's plates are not simply those which would catch the eye of the artist or the gardener; rather, they are inserted because they are diagnostic. They determine the character of the plant, its species and genus. Thus, they increase the informational value of the illustration substantially.

The *Hortus Cliffortianus* reflects Linnaeus's new approach to the organization of botany. It is important to realize that Linnaeus's most salient character-

PLATE 41: Jan Wandelaar (1690–1759), after Georg Dionysus Ehret (1708–1770), *Collinsonia*, 1737–8. Engraving. From Carolus Linnaeus, *Hortus Cliffortianus* (Amsterdam, 1737–8), pl. V.

istic was his taste for order, and that one of the great problems in eighteenth-century natural history was the chaos that reigned in the naming and description of plants. In his *Philosophia Botanica,* published in 1751, Linnaeus relates, in a series of aphorisms, his understanding of the history of botany. His cryptic phrase, 'America 1492,' encapsulates Linnaeus's realization that the upsurge in botany in the sixteenth century was the direct result of colonial expansion and the increase in the number of new and exotic plants. That increase had not slowed down in the early eighteenth century, and gardeners such as Clifford, Peter Collinson, and John Fothergill in England, and Alexander Garden in America, were always on the lookout for rarities. How were these to be named, to be systematized? Would new plants have to remain *herbae nudae,* unconnected with each other? What was the basis by which plants could be grouped and known? Linnaeus's system for plant identification was the equivalent of 'Ariadne's clew for botany without which all [the kingdom of plants] is chaos.' With his clew, Linnaeus could unravel the most difficult problems: 'Take for instance an unknown plant from the Indies, a "botanophile" will look through descriptions, illustrations, all indexes, but he will not find its name except by pure chance; a taxonomist, however, will soon establish the genus, whether the old one or the new.'[18] Linnaeus was the taxonomist, whose system was based on a logical method by which one could at a glance determine the true character, the essence of the plant. Linnaeus dismissed as irrelevant other attempts at systems, which he referred to as 'heterodox.' These included botanists who classified plants by generic name (the alphabetics), by root (the rhizotomi), shape of leaf (the phyllophiles), by flowering time (the chroniclers), by habitat (the topophiles), by medicinal properties (the empiricists), and by the order of the traditional pharmacopoeia (the pharmacists). The correct or orthodox way to classify plants was by their fructification. Not only is it an essential part of the plant, it is, as Stafleu notes, morphologically complex and amenable to simple numerical classification.[19] Linnaeus lists the seven parts of the fructification: the calyx, the corolla, the stamen, the pistil, the pericarp, the seed, and the receptacle. These are the diagnostic parts of the plant. It is these which must be included in any illustration and which will then allow determination of genus. They mark the plant and give it its character.

The Linnaean system was not a natural system. As Linnaeus states in the *Philosophia,* botany is 'that part of the natural sciences by the help of which one obtains happily and easily a knowledge of plants and by which one remembers this knowledge.' It is a device to register, remember, to store and retrieve.[20] All botanists should remember the genera, which are fixed, and thus the recognition of a new plant is as simple as counting the parts of its fructification and assigning it to an existing genus, or creating a new one. Linnaeus used his new method to bring order to the confusion of botanic names. Where Cornut had identified the columbine as *Aquilegia pumila praecox canadensis*, Linnaeus would usually permit only two names, a generic and a specific. His aim was to establish a series of euphonic, easily remembered generic names, based (for the most part) on Latin and Greek roots. He wrote that his 'Genera Plantarum is a work such as no one before him had done, which describes all parts of the fructification in the species exactly and thereon makes the characters, so that a genus not described according to Linnaeus's method is incompletely known.'[21] Not everyone agreed with Linnaeus, accusing him of wanting to be a second Adam, but his system proved useful, and was championed by the British botanist and first president of the Linnaean Society, James Edward Smith (1759–1828), as the best artificial system to be devised until a natural system could be implemented. Without this simple system, Smith asserted, the 'celebrated DR. GARDEN who studied by it, assured me that ... he would probably have given up the science in despair, had not the works of LINNAEUS fallen in his way.'[22]

The Linnaean system was not universally accepted, but Linnaeus's idea of using binomial names for plants and animals rapidly caught on. Linnaeus was the eighteenth-century pandect. Like Gesner, he had his contacts throughout Europe and beyond, depending on a veritable army of students and correspondents to send him dried plants, skins, books, and drawings. His pupil Pehr Kalm, for example, whose account of his travels in North America from 1748–1751 were translated into English as early as 1751, supplied him with many North American plants. In all, he developed binomial and internationally usable names for 4,400 species of animals and 7,700 species of plants. England became a champion of the Linnaean system, and its application had a great impact on the work of English naturalists and explorers. In England, Linnaeus had long been in contact with Peter Collinson, John Fothergill, and John Ellis, though his system was not taught at the universities until 1762. George Edwards included the Linnaean classifications in later editions of his work in an attached catalogue of the birds made by Linnaeus himself, dated

'Upsala 1775.' John Ellis wrote to Linnaeus, on the occasion of Cook's expedition in the *Endeavour*, that 'no people ever went to sea better fitted out for the purpose of Natural History. They have got a fine library of Natural History ... They have two painters and draughtsmen ... All this is owing to you and your writings.'[23] His *Systema Vegetabilium* was translated into English in 1783 by a 'Botanical Society at Lichfield,' and it is worth quoting a passage in full which shows the supposed ease by which Linnaeus's system, with its key, could in fact identify plants:

> The BOTANIST, in following the *Classifications*, is [led] to the named *Genus* by the *Characters* of the displayed plant or flower; to the appellation of the *Species* by the *Differences* of the Larva or herb; and thence to its *Synonymies*; from these to the *Authors*, and thence to *every thing*, which has come to us from our ancestors on the subject. Thus the plant itself tells us its Name and its History amid such a multitude of species, and of individuals; this is the great purpose of Botany, the invention of the present age, to the completion of which all true Botanists will contribute their labour.

The passage goes on to note that 'true BOTANISTS will labour to increase this lovely Science: will construct Fundamental DESCRIPTIONS in characteristic words ...' and 'Will add FIGURES, if they are able, which represent the perfect Plant ...'[24]

For Linnaeus, the work of the illustrator was significant, and he notes in *Philosophia Botanica* that 'a painter, an engraver and a botanist are all equally necessary to produce a good illustration; if one of them goes wrong, the illustration will be wrong in some respect. Hence botanists who have practiced the arts of painting and engraving along with botany have left us the most outstanding illustrations.'[25] He also observes that all illustrations should be of natural size, and, as noted above in the description of the *Hortus*, placing a six-foot plant on a folio sheet involved some manipulation by the artist. It should be remarked, however, that to the botanist, the herbarium was primary. Linnaeus stated that 'a herbarium is better than any illustration; every botanist should make one.'[26] Unfortunately, herbarium specimens were subject to the same depradations of insects, mould, and pests as stuffed skins and dried birds. Nikolaus Joseph Jacquin (1727–1817) collected West Indian and Central American plants, but, in the 1780s, 'ants damaged Jacquin's herbarium material, and he therefore supplemented his descriptions and notes on the new species with water-colour drawings. These accordingly are the equivalent of type-specimens.'[27] It made sense, then, that botanists should attempt to protect their specimens not only in herbaria, but also in illustrations, which, when engraved and coloured, could be widely disseminated. In Blunt's biography of Linnaeus, there are two plates which reveal two aspects of the use of engravings of plants. In a painting by Jacob de Wit (1695–1754), now in the Orangery at Uppsala, entitled *Three People Displaying a copy of Linnaeus' 'Hortus Cliffortianus,'* two older gentlemen and a young women look intently at one of the plates. A herbarium specimen lies on the table, and one of the older men appears to be explaining something to the others. The woman is well dressed, and is wearing pearl jewellery, an indication that the purchase of a book like the *Hortus* was not for the impecunious. A second plate shows photographs of two rooms in Linnaeus's house at Hammarby, in Sweden. The walls are literally papered with coloured engravings, folio size, of plants. A contemporary observer visiting Hammarby in 1765 noted that 'the walls of Linnaeus's rooms at Hammarby were covered with botanical prints taken from the finest volumes of Sloane, Ehret, etc., and pasted on them so that they looked just like wall-paper ...'[28]

It is no surprise, then, that Linnaeus also accounted Ehret's flowers a miracle of the century, and that Ehret was almost continuously employed throughout his life by collectors and naturalists. Georg Dionysus Ehret was the most famous flower painter of his age and, like Edwards, worked for a number of the great European collectors, including the Margrave Karl III Wilhelm of Baden-Durlach, whose gardens at Karlsruhe were famous; Dr C.J. Trew, of Nuremberg, Ehret's lifelong friend and patron; and Joseph Banks. Banks's patronage revealed his embrace of Linnaean methods, encouraged by his first librarian and curator, Daniel Solander (1733–1782), one of Linnaeus's pupils sent to Britain in 1760 at the request of Collinson and Ellis. Ehret was hired by Banks to prepare a number of watercolours from the dried specimens he had brought back from his 1766 expedition to Newfoundland and Labrador. Ehret had been identified with Linnaeus and his method since the publication of the *Hortus* in 1737. In addition, Ehret had published a 'Tabella' of the Linnaean system, which he had learned from Linnaeus: 'Linnaeus and I were the best of friends: he showed me his new method of examining the stamens, which I easily understood, and privately resolved to

PLATE 42: Georg Dionysus Ehret (1708–1770), Rhodora (*Rhododendron canadense*), 1767. Watercolour on vellum. The rhodora was collected at Inglie in Newfoundland.

bring out a Tabella of it.' He published the plate in Leyden, and its sale proved quite profitable, 'for I sold it at 2 Dutch gulden a piece and almost all the botanists in Holland bought it of me.'[29] For Banks, Ehret painted twenty-three of the Newfoundland plants on vellum, and, like all Ehret's work they are delicate, lively, and true (plate 42). They differ fundamentally, however, from the gorgeous tropical growths of Merian, and the flat, if exact designs of Catesby. The difference is in their value as information, and this value stems from the adoption of Linnean classification and the delineation of the characters.

The Universal Language

Plants were not the only living things to have a 'character.' Linnaeus had published the *Systema Naturae* in 1735, outlining a zoological classification system. The tenth edition, published in 1758, provides binomial names for all known animal species, and is internationally accepted as the starting-point for modern zoological nomenclature. William Hunter included 'Linnaeus Description' in his unpublished account of the moose, noting that 'the Characteristics given by Linnaeus will apply to the Orignal ...'[30] Hunter's own description (originally in Latin) provides a good example of the 'character' for an animal: 'Deer with long horns, short neck, big head, gibbous nose. Guttural caruncle slender, mane erect, shaggy and dense; almost no tail.'[31] While this description might act as a key to distinguishing the moose as a specimen from other specimens of deer, elk, or caribou, it is not of great help in identifying an actual animal if the naturalist had never seen one. Thus, Hunter had commissioned Stubbs to provide the painting in 1770, and, when he sought to compare a second moose with the first, he and Daniel Solander and several others went to visit the Duke of Richmond's recent arrival on 4 October 1773, 'carrying with us Mr. Stubbs's picture.' It is hard for contemporary North Americans to imagine not carrying around a mental picture of a moose and being able to recognize it instantly when the real animal appears, but in eighteenth-century England not only were living moose a rarity, but their images were just as rare, and thus a group of naturalists was forced to carry along an oil portrait in order to ascertain the differences, if any, between the first and second moose. Hunter and his party spent some time measuring the second moose and observing his behaviour, noting the differences. In July 1772, Hunter had another opportunity to see a male moose and recorded that 'as far as I could distinguish with the Eye there was no material difference between him and this of the D. of Richmond ...'[32]

Hunter was well aware of the importance of accurate illustration. He included in his essay a section called 'Reflections' in which he discussed the work of other previous authors on the nature of the 'orignal.' He opens the discussion by stating the problems with textual description alone: 'This subject shews us how vague our Ideas of natural productions are when made out by descriptions only. Any person who has seen this Animal, or even this picture only of him, would know him again at one glance if he should meet with him in any part of the world.' Hunter credited illustration as a universal language and it is worth quoting his 'Introduction to Reflexions' in full:

> All Men educated to study lament the variety and confusion of Tongues; & while they are sensible of the advantages which would arise from one language being universal, they see that they must ever regret the want of that blessing. Yet for many purposes, especially in the Arts, and in natural History there is a language which is both easily acquired, &, tho not so copious, is more expressive than any language in the world, and at the same time so plain that the unlearned as well as the learned, understand it at first sight: I mean the art of drawing. What pity it is that it had not been sooner introduced, and more generally used.
>
> As descriptions in this Language are so expressive, so precise and well determined, they have more credit than descriptions in common language; especially too as they are all presumed to be taken from the Life. Thence impositions and misrepresentations are more unpardonable in this way: for in as much as they have more credit, they do more mischief.[33]

Several years earlier, George Edwards had written in a similar vein concerning the value of illustration in natural history in the preface to *Gleanings*:

> Where accurate figures are given, much pains may be spared in verbal descriptions, by referring to the figures, which lineally describe the minutest parts of them, and such as would be very doubtfully understood by words only: indeed, the figures in this work cannot be deemed perfect without their descriptions, and the descriptions as nothing without the figures; ...

> The prints through this work, when truly illuminated, may be considered as a book legible to people of all nations and languages, whether learned or illiterate: real representations of animals &c. properly delineated and coloured, are characters that all nations are taught by nature to understand; and in many respects, good figures from nature surpass the best verbal descriptions.[34]

Art is the universal language, and careful drawings after life are made in the image of nature itself, universally accessible to all. Those who would study the book of nature do not require Latin to read the characters, to determine the essence of the living thing; it can be read in line and paint. Thirty years later, in 1796, Joseph Banks, now president of the Royal Society, wrote an appreciation of Franz Bauer's *Delineations of Exotick Plants cultivated in the Royal gardens at Kew*, and authoritatively pronounced that 'every Botanist will agree, when he has examined the plates with attention, that it would have been a useless task to have compiled, and a superfluous expence to have printed, any kind of explanation concerning them; each figure is intended to answer itself every question a Botanist can wish to ask respecting the structure of the plant it represents.'[35] Despite his own admission that the reproduction of his works could never truly imitate nature, Edwards, and naturalists such as Hunter and Banks, were convinced that coloured figures could be executed in a style whose conventions were those of nature itself. The manuscript drawing, the medium of exchange between naturalists for centuries, had attained a new scientific authority as a result of two sometimes conflicting ideas about the nature of scientific truth, the first allied to the importance of observation, and the second to the belief in essential reality, or the revelation of the natural order of the world.

Truth and Observation

Linnaeus ended both his *Fundamenta* of 1736 and his *Philosophia Botanica* of 1751 with these words: 'In natural science the principles of truth ought to be confirmed by observation.'[36] George Edwards was, throughout his works, concerned with the nature of truth and falsehood, particularly in the description of animals. 'Man,' he wrote 'ought to set before his intellectual mind the ideas of truth and falsehood, and endeavour to find out, in the most strict and

PLATE 43: George Edwards (1694–1773), ***The Porcupine from Hudsons Bay***, 1742. Etching, hand-coloured. From George Edwards, *A Natural History of Uncommon Birds*, part 1, plate 52. Edwards drew the porcupine from a specimen in the collection of Sir Hans Sloane, who had received it from Dr R.M. Massey at Stepney, who in turn had been sent the specimen by Alexander Light from Hudson Bay.

absolute sense what they are; and when he hath found them he ought to govern all his actions by the former, and avoid the latter but it is exceedingly hard to discover what truth is in a world of falsehood and controversy, where all of us suck in error with our mother's milk ... it is a firm and fixed article of my private faith, that God has given us our senses as touchstone of truth.'[37] Edwards's meticulous textual descriptions and his insistence on reporting only first-hand observation reflect this private faith in the value of sensory evidence. He had harsh words for the showmen, 'who shew foreign birds and beasts.' In a tradition that reaches far back into the world of the bestiaries and is continued in the works of the sixteenth-century pandects, Edwards maintains that, 'to make them seem more rare, [showmen] often pretend them to be natives of places very distant and unknown ...; and, to strike us with surprize, they pretend that to be fierce, savage, and untameable creature, which in its real nature is very gentle and harmless.'[38] Edwards give as an example the slow-moving porcupine, which some claim to be fierce and able to shoot its quills, which Edwards repudiates, illustrating a quill of its natural length as well as a detail of the barbed end (plate 43). (Edwards's reputation for accurate observation encouraged imitation. His porcupine, engraved in 1742, was copied by Henry Ellis in *A Voyage to Hudson's Bay by the Dobbs Galley and California ...*, published in London in 1748 [plate 44]. It was also copied in reverse by Moses Harris on *A Plan of the Harbour of Chebucto and Town of Halifax* published in London in the February 1750 issue of *The Gentleman's Magazine* [plate 45].)[39]

The emphasis on observation of the thing itself was shared by other illustrators of natural history. Eleazar Albin, for all that Edwards criticized his birds, was an acute observer of other species, especially insects and spiders, on which he published *A Natural History of Spiders and Other Curious Insects* in 1736. He noted in his preface 'To the Reader,' that

> But as the world, in intricacies of nature are commonly dubious of the facts related, I assure the reader that the accounts I have here given have been from my own ocular observation, having made it my particular care to be extraordinary exact in the collections and observations I have made concerning this Insect.
>
> ...
>
> Each particular species of this little insect (of which I have with great pains and labour collected near 200) is here represented, and (for such as desire it) beautified in its proper colours, that the reader may have his curiosity satisfied at one view, by what has taken me up a great deal of time and labour, to put so exactly together.[40]

Paillou included the measurements of birds on his watercolours for Taylor White, who insisted they be painted life-size. In 1737, Ehret travelled every day for several weeks to Sir Charles Wager's garden in Fulham to observe each stage in the flowering of the *Magnolia grandiflora* (included in Catesby's *The Natural History of Carolina ...*). Edwards accepted William Bartram's drawing of a marsh hawk, executed in Pennsylvania, because, though he had not seen

PLATE 44: After George Edwards (1694–1773), ***The Porcupine and The Quick Hatch, or Wolverene***, 1748. Engraving. From Henry Ellis (1721–1806), *A Voyage to Hudson's Bay by the Dobbs Galley and California* ... (London, 1748), facing page 42.

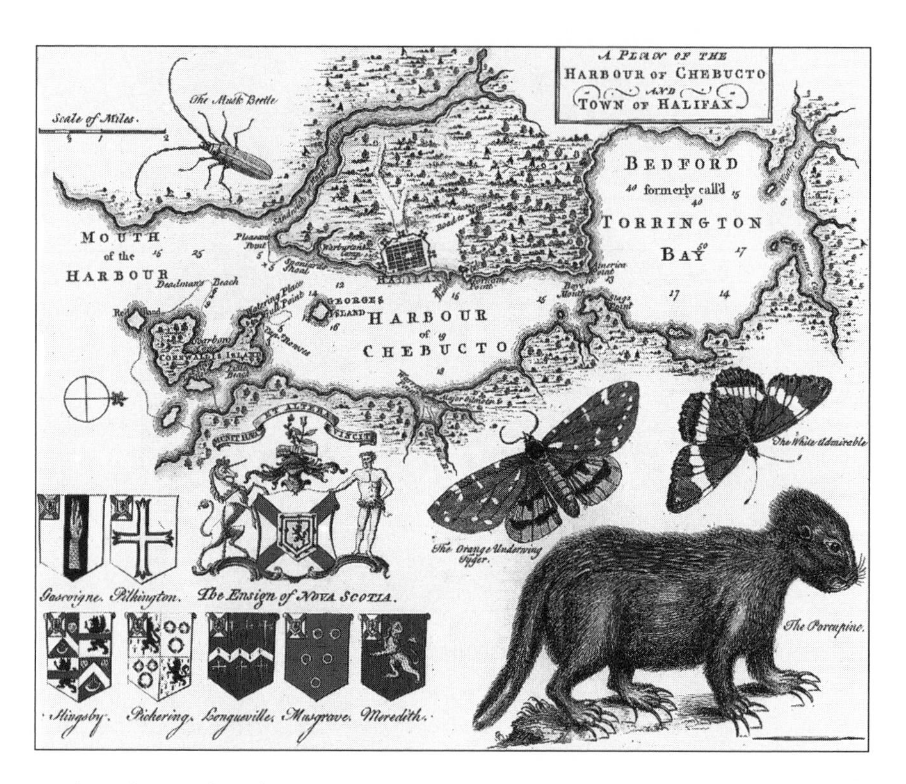

PLATE 45: Attributed to Moses Harris (1730–1788?), ***A Plan of the Harbour of Chebucto and Town of Halifax***, 1749. Engraving. From *The Gentleman's Magazine* (London, February 1750). Moses Harris was presumably the author of *An Exposition of English Insects ...* (1776), which featured exquisite hand-coloured engravings of European insects. On this so-called porcupine map, Harris copied the porcupine from Edwards, but it is likely that the butterflies and beetle are his own work.

the bird itself, 'I have great reason to think Mr. Bartram is very correct in his drawing, and exact in his colouring, having compared many of his drawings with his natural subjects, and found a very good agreement between them.'[41]

For the naturalists, 'ocular observation,' relying, as Edwards had suggested, on the touchstone of the senses, had completely replaced reliance on ideal images, memory images, translations of textual description, and copying of iconic images. The constant checking against nature, which is so evident in the works of Edwards and Ehret, meant that, for the first time, naturalists had access to a large body of reliable pictorial evidence about plants and animals. It should be re-emphasized, however, that drawing after nature did not guarantee an image of the living animal. Drawing from the specimen guaranteed only a likeness of the specimen, and, therefore, Edwards's figure of the porcupine, distorted to contemporary eyes, is a reliable scientific illustration for the naturalists of the day, because it meets their requirements for information. The information that many naturalists required after mid-century was that which would allow them to determine the character of the plant or animal according to the Linnaean system, and that system was based on appearance as the key classificatory factor.

The Linnaean system, as was apparent to many naturalists at the time, is not a natural system. Natural systems classify plants and animals according to their affinities as observed in nature. The burgeoning importation of new species into Europe in the eighteenth century made the possibility of developing true natural systems seem highly unlikely. An artificial system could at least provide some order until natural affinities had been determined. Linnaeus had to choose an element of the plant or animal that would allow rapid and efficient categorization. He rejected the classifications based on the internal properties of the plants, such as cross-sections of roots, or inherent medicinal properties; in forming specific diagnoses, he also rejected those that were 'inconstant, purely quantitative, ecological or geographical ... He banned as misleading, indefinite or unhelpful characteristics those of size, resemblance to other genera, locality, time of development, colour, smell, taste, uses, sex, monstrosities, hairiness, duration and increase in number of parts.' Similarly, in regard to animal species, he rejected many of the fortuitous characteristics that had been used by earlier authors, such as usefulness to man, but 'made much use of colour, particularly in birds and insects, and also of habitats and hosts in insects.'[42] The characteristics which he declared most salient were, in fact, those most amenable to observation in collections of specimens or in illustrations. The artificial nature of Linnaeus's system was to a certain degree dependent upon the artificiality of his materials.

While Linnaeus did maintain a garden, he primarily relied for information about exotic species of plants on preserved and dried specimens. For example, he prepared his *Flora Zeylanica* (Flora of Ceylon, 1747) on a collection of dried specimens made by Paul Hermann between 1670 and 1677. He was delighted with the gift from the King of Sweden in 1775 of 'sixteen great chests containing plants preserved in spirits of wine – just as they grow, with their flowers and fruit. There they were, so well preserved that they looked as though they had just been gathered ... There is all the difference in the world between seeing plants pressed and dried and seeing them as they grow.'[43] Similarly he classified Edwards's birds on the basis of their coloured figures, which, as we have discussed, were primarily prepared from dried or stuffed skins. Appearance, then, not structure, behaviour, or even reproductive exclusivity, was the key classificatory factor for Linnaeus's artificial system. It is interesting to note here Martin Kemp's assertion 'that there were special kinds of affinity between the central intellectual and observational concerns in the visual arts and sciences in Europe from the Renaissance to the nineteenth century. The affinities centred upon a belief that the direct study of nature through the faculty of vision was essential if the rules underlying the structure of the world were to be understood.'[44] While Linnaeus developed his system to be almost an *aide-mémoire* for the botanist, he did not regard it as wholly distinct from nature. For Linnaeus, the genus was a fixed part of the first Creation. Stafleu has characterized Linnaean thought as Aristotelian and Thomistic, and there did exist for Linnaeus a knowable 'essence' of an ideal plant.[45] This was the genus, and the description of the character reveals the essence of the genus. By selecting the fructification, whose function was so important, indeed essential, in maintaining the type, Linnaeus was reflecting the natural order of the world. By describing and figuring the actual plants in the Linnaean manner, the naturalist is providing not just a realistic representation of a living thing, but also a representation of its very essence as expressed through its character. The illustration is not simply a counterfeit of appearance; rather, it reveals a deeper truth about the object and its place in the natural order. The accumulation of accurate descriptions and representations of living things is an enterprise whose objective is the revelation of the order of Creation. Understanding the essence is, in a sense, seeing into the mind of God.

This connection between visual reality and the true nature of the world underlies much of what might appear to us as the peculiarly moral and religious language of a number of eighteenth-century naturalists. Edwards, as we have already noted, dedicated his first work to God, and believed that God had given him his senses as touchstones of truth. His works must be as exact as humanly possible, for in illustrating the birds and animals he is attempting to understand the mind of the Creator. This recognition of the connection between illustration and revealed truth explains Hunter's almost religious zeal in condemning those who would misrepresent animals (see above), or Thomas Pennant's ardent desire for a representation of Banks's Patagonian penguin: 'Let me hope that the Patagonian Penguin had set for its picture, that Mr. Brooks' Percnopteru will not depart this life without having its image preserved to be transmitted to posterity by Mr. Paillou's pencil; that the image of these and many others may for the benefit of the curious and making of proselytes to our divine science be multiplied by engraving and that we may with unabated zeal pursue the path we have begun by our four plates.' Impatient for receipt of the image, Pennant wrote again to Banks: 'Is your Penguin drawn? I dream, I rave of it?'[46] Preserving images for posterity, hoping for proselytes, seeing Edwards's birds and Ehret's flowers as miracles, dedicating books of illustrations to God – all these are evidence of the importance which the naturalists and their artists assigned to the depiction of the natural world. Their work was not simply the illustration of collections, it was the illustration of Creation, and through correct and exact illustration they might come closer to the essence of things themselves. This sense of purpose helps to explain the extraordinary petition written by George Edwards and included in the *Memoirs*:

> My petition to God (if petitions to God are not presumptuous) is, that he would remove from me all desire of pursuing Natural History, or any other study ... What my condition may be in futurity is known only to the wise disperser of all things; yet my present desires are (perhaps vain and inconsistent with the nature of things!) that I may become an intelligent spirit, void of gross matter, gravity and levity, endowed with a voluntary motive power, either to pierce infinitely into boundless etherial [*sic*] space, or into solid bodies; to see and know, how the parts of the great Universe are connected with each other, and by what amazing mechanism they are put and kept in regular, and perpetual motion.[47]

The naturalist-artists and authors of the eighteenth century relied on the visible appearance of things and, through careful observation and accurate description and depiction, saw, as had their predecessors who sought the signatures and marks in animals and plants, the essential characteristics that revealed hidden affinities and, thus, the order of the world. It would remain for Charles Darwin, another British naturalist, nearly a century later, to unravel the amazing mechanism of the historical process of selection by which the parts of the living universe were connected one with another.

seven

A Country Observed

Observation was one of the watchwords of the eighteenth century. In the English-speaking world, James Isham, George Edwards's correspondent from Hudson Bay, wrote his *Observations* in 1743, while John Bartram, the great American collector, published his *Observations on the Inhabitants, Climate, Soil, Rivers, Productions, Animals*, ... in 1751. We have already noted Albin's assurances to his readers of his skill in 'ocular observation.' Daines Barrington (1727–1800), correspondent of Gilbert White's, urged naturalists to keep a journal where they could record the 'many other particulars [that] daily offer themselves to the observer' in the hope that 'from many such journals kept in different parts of the kingdom, perhaps the very best and accurate materials for a General Natural History of Great Britain may in time be expected ...'[1] Barrington, George Edwards, Joseph Banks, Thomas Pennant, and other members of the Royal Society were imbued with the passionate desire for observable facts that had been at the heart of the Society since its foundation. The Society's stated purpose was 'to study *Nature* rather than *Books*, and from the Observations, made of the *Phaenomena* and Effects she presents, to compose such a History of Her, as may hereafter serve to build a Solid and Useful Philosophy upon ...'[2]

The observation of nature that the Society espoused confronted, in true Baconian fashion, the reliance on authority which had been the mainstay of early naturalists. The observations now required were of a different order from the casual notes of the learned Dr Cay on the Lumpus, quoted in chapter 6. Observations were to be scientific, measurable, and accurate. The standard was not the authority of the scholar, but the objectivity of the observer. This new reliance on objective observation broadened the scientific community, allowing it to incorporate within its bounds any literate correspondent with a taste for natural history and proven reliability. From the 1660s on, the Royal Society provided sets of 'Directions or Inquiries' for seamen, settlers, and others travelling to foreign parts. They requested observations on everything, from tides and weather to height of trees, quality of soil, types of crops, and kinds of animals. They demanded confirmation of travellers' tales and unusual stories, and provided detailed instructions on how to collect data, and even how to construct appropriate collecting tools and instruments. This zeal for data collection diminished little with time, fuelled as it was not only by the desire for new knowledge, but by the conviction that knowledge of the observable world would lead to knowledge of God. If the world and all its creatures were indeed the creation of a Divine Workman, then surely close examination of things would reveal more clearly than any other method the mind of the Creator. Robert Hooke, one of the Society's founders and author of *Micrographia* (1664), believed that he could use his microscope to see further into the heart of things, to recover an Adamic ability to read the 'micrographia' inscribed by God.[3] Fifty years later, William Derham, writing in *Physico-Theology* (1711–12), reiterated the new agenda for natural history: 'Let us ransack the globe, let us with the greatest accuracy inspect every part thereof, search out the innermost secrets of any of the creatures, let us examine them with our gauges ... pry into them with all our microscopes and most exquisite instruments, till we find them to bear testimony to their infinite workman.'[4]

The broadening of the circle of observers and the development of objective measures for observation had enormous impacts on the compilation of the natural history of North America, giving new authority and meaning to the work of the resident naturalists of the New World, particularly in relation to zoology. With the notable exception of Banks, the majority of the European naturalists stayed firmly in their cabinets, at the most making local excursions to remoter areas of Europe, such as Thomas Pennant's trip to Scotland or Linnaeus's Lapland journey. They relied on their correspondents to 'ransack the globe,' and established far-flung networks of collectors and informants who shared their enthusiasms and their scientific method. These observers in foreign parts fell into four groups: residents, like John and William Bartram;

military or naval officers on a tour of duty, like Thomas Davies (*ca* 1737–1812); travellers and explorers, like Linnaeus's pupil Pehr Kalm (1716–1779) or Samuel Hearne (1745–1792); and resident employees of the Hudson's Bay Company, like Isham, and later Andrew Graham (d. 1815) and Thomas Hutchins (d. 1790). Some of these observers were highly trained, like Kalm; others, like Thomas Davies, were gifted amateur naturalists; while still others, particularly the Hudson's Bay Company factors who undertook their collecting activities at the request of the Company and the Royal Society, found in them both a diversion and a sense of purpose that helped maintain their spirits during the long northern winters. James Isham, the factor who, on his return to England in 1748, presented George Edwards with some thirty specimens, wrote to the deputy governor and commissioners of the Hudson's Bay Company that he had undertaken his 'Observations' to promote his mental health: 'Being in a Disconsolate part of the world, where there is Little conversation or Divertisment to be had, I was dubious of that too common Malady the Vapour's, which is frequent the forerunner of other Distempers, therefore to prevent such if possable, I have in cold Days and Long winter Nights, amusd. my self with the following Observations ...'[5] The care with which these resident naturalists made their observations is obvious in their journals and, in the case of Davies, in his meticulous watercolour drawings. Their dedicated inspection, which involved making their own observations, querying Native hunters and employees, examining stomach contents, and, in the case of Hearne, dissecting specimens and examining them under his 'excellent microscope'[6] (as Derham had advised), meant that, even far from the centres of science, they could make meaningful contributions to the marvellous enterprise of natural history. Their contributions changed not only the content, but the direction of natural history at the end of the eighteenth century, leading to a new appreciation for field observation, and to a new kind of science based, not on the cabinet or the herbarium, but on nature observed.

No extravagant wonders ...

Pehr, or Peter, Kalm undertook a trip to North America between 1748 and 1751. Kalm was one of Linnaeus's star pupils, and the source of much of his master's North American material. His account of the expedition was published in Swedish, but a letter in English to John Bartram about Kalm's trip to Niagara Falls was appended to Bartram's 1751 *Observations*. The letter, dated 1750, was also printed in *The Gentleman's Magazine*, with an accompanying illustration that appeared in a subsequent issue. Kalm assures his readers that he has provided 'a short but exact description of this famous Niagara cataract.' He goes on to protest his veracity, stating 'you may depend on the truth of what I write. You must excuse me if you find in my account no extravagant wonders. I cannot make nature otherwise than I find it. I had rather it should be said of me in time to come, that I related things as they were, and that all is found to agree with my Description; than to be esteemed a false Relater.'[7] Kalm approached the falls with some scepticism: 'I had read formerly almost all the authors that have wrote any thing about this Fall; and the last year in *Canada*, had made so many enquiries about it, that I thought I had a pretty good Idea of it ... But as I found by experience in my other travels, that very few observe nature's works with accuracy, or report the truth precisely, I cannot now be entirely satisfied without seeing with my own eyes whenever 'tis in my power ...'[8] Once he arrives at Niagara Falls, Kalm reports his observations with careful attention to objective description: 'Half an hour past 10 in the morning we came to the great Fall, which I found as follows. to the river (or rather strait,) runs here from S.S.E. to N.N.W. and the rocks of the great Fall crosses it, not in a right line; but forming almost the figure of a semicircle or horseshoe ...' He goes on to describe the division of the waters by the island at the brink, and their swiftness, then at last gives in to a touch of amazement: 'When all this water comes to the very Fall, there it throws itself down perpendicular! It is beyond all belief the surprize when you see this! I cannot with words express how amazing it is!' He disputes Father Hennepin's measurements, however, calling him '*The great Liar*,' and notes that, for his part, he is 'not fond of the *Marvellous*, I like to see things just as they are, and so to relate them.'[9] He does, however, report some of the observations of 'the *French* gentlemen,' who relate that 'when birds come flying into this fog or smoak of the fall, they fall down and perish in the Water.' Others maintain that it is rather water birds like swans, geese, ducks, and teal that float along the water until 'they are driven down the precipice.' The numbers of these dead birds are so great that, 'in the months of *September* and *October*, such abundant quantities of dead waterfowl are found every morning below the Fall, on the shore, that the garrison of the fort for a long time live chiefly upon them ...'[10] Kalm also relates the story of two Native people who were stranded on the island in the midst of the falls and attempted to escape by building a wooden ladder down to the river.

Despite his care in preparing an accurate textual description, the truth value of Kalm's account is negated by the illustration that follows it in *The Gentleman's Magazine*. This figure is yet another re-engraving of a version[11] of Hennepin's view of Niagara, almost a century and a half after it was first published. Thus, despite Kalm's dismissal of the account of the 'great Liar,' the image of Niagara in Hennepin's book, which even a cursory reading of Kalm would show was inaccurate, remained the one most available to Europeans. Several details have, however, been added to the original or changed by the engraver, to correspond more plainly with Kalm's account. There is an attempt as well to make the illustration more 'scientific' by the addition of a letter code, designating specific details (like the codes on Ehret's plant drawings). For example, 'a' indicates 'The Place where a Piece of Rock was broken from which while standing turned the Water obliquely across ye Fall as in Popple's Map'; 'b' shows 'Two Men passing over ye east Stream with Staves'; and 'c,' 'The Indians reascending their Ladder.' As in the earlier engraving, Europeans stand to the left, amazed, while on the right a line of Native people toil over the portage which Kalm describes. In keeping with Kalm's remarks on the height of the falls, some attempt has been made to change the scale of the image, so that the falls appear less lofty. The fir trees, rather than standing in straight lines, as in the Hennepin engraving, are now bent at an angle, a convention indicating wilderness. The veracity of Kalm's first-hand observation of Niagara was, however, seriously diminished by the distortions necessitated by reusing an image that was not created by an observer on the spot. For example, the image shows a second stream entering the falls from the left, which is not in the Hennepin image but seems to have been added by Henry Popple in his 1733 version. Whereas Edwards could claim that the figure and the description are in essence one and the same, the illustration for Kalm's account matches the textual description only in the ascription of what are essentially narrative details, where an improbable ladder can stand for a tale of near disaster, a group of tiny figures for an example of ingenuity, and a geometric flock of emblematic birds for a natural phenomenon. Despite Kalm's attempts to introduce a scientific objectivity into his description, the accompanying image is inaccurate and in essence untrue, a schematic representing the *idea* of the falls rather than an accurately observed depiction of them.[12]

The understanding of what was required in the accurate depiction of landscape was rapidly changing, however, with the work of the English watercolourists and draughtsmen who were now being called upon to picture the homes and estates of the landed gentry, and the coastlines and fortifications of foreign lands with some degree of verisimilitude. Over a decade after Kalm's article was printed, Thomas Davies issued a portfolio of engravings of six North American waterfalls, including an engraving of Niagara very different from the image presented in *The Gentleman's Magazine*. Davies's Niagara is the result of personal observation, and is executed with the kind of fidelity to things as they are that was part of Davies's training as a topographical artist.

Thomas Davies was a graduate of the Royal Military Academy at Woolwich, where the development of some skill in draughtsmanship was part of the curriculum. Military officials had long been aware of the importance of ensuring that their graduates be able to render plans, and sketch maps and some descriptive views. The inspector general of the Marlow Military College stated the purpose of this training succinctly: 'Everything which is put down in writing of necessity takes on some colour from the opinion of the writer. A sketch map allows of no opinion.'[13] Davies was likely trained in painting by Gamaliel Massot, a relatively obscure artist who was succeeded in 1768 by Paul Sandby (1731–1809), known as 'the Father of the English School of Watercolour,'[14] with whom Davies likely came in contact during his periodic returns to Woolwich. At the request of his superiors, Davies produced views of fortifications, plans and elevations of vessels, schematic figures of artillery formations, and records of battles. He also painted to please himself and to record both the landscape and the wildlife of the New World. Waterfalls were a particular passion, as they had been with many North American travellers. So much of transport was by river, and rapids and cascades were objects to be named and remembered both for the danger or discomfort often involved in their passage, and for their punctuation of an often tedious voyage. For Davies and other painters of his time, however, they were also one of the key elements in the picturesque landscape. The picturesque eye demanded a certain 'roughness,' and, according to William Gilpin, the English author who helped to define its qualities, artists and aesthetes could seek the picturesque 'among all the ingredients of landscape – trees – rocks – broken-grounds – woods – rivers – lakes – plains – vallies – mountains – and distances.'[15] The waterfall epitomized this aesthetic roughness, and between 1760, after the capture of Montreal, and 1766, Davies travelled extensively through Canada and the northeastern United States, painting the six waterfalls he later had engraved in England between 1763 and 1768. He made the sketch of Niagara that served as the model for the engraving in 1762, and, about four years later, he painted two additional watercolours of the great falls.

All of Davies's views are remarkable, providing for the first time an accurate representation of the falls. The 1762 view is entitled *An East View of the Great Cataract of Niagara*, and includes the information 'The Perpendicular height of the Fall 162 feet / Done on the spot by Thomas Davies Capt Royal Artillery / The Variety of Colours in the Woods shew the true Nature of the Country.' Dismissed by Hubbard as 'too perfunctory to be of much artistic interest,'[16] this view is nevertheless of great interest from the perspective of the understanding of the illustration as information. Not only are the falls depicted in their correct horseshoe configuration, but their height is given with reasonable accuracy, and the woods are shown in their autumn colours. Hubbard suggests this may be the first appearance in art of the colours of the Canadian mixed forest in the fall, and Davies feels obliged to add a note, perhaps in case his viewers feel the colour is imaginary, that 'The variety of Colours in the Woods shew the true Nature of the Country.'[17] The engraving by Ignace Fougeron (fl. *ca* 1760–1768) is surprisingly faithful to the original watercolour sketch, and distorts very little Davies's representation (plate 46). Two Native people in elaborate dress stand to the right, on the edge of the river, gesturing perhaps at the geometrically precise rainbow that arches over the river. The 'smoak,' which Kalm described, rises over the falls, and the trees come down close to the water's edge and form a mixed forest with some conifers, typical of southern Ontario. The surface of the river shows strange flat rocks, rather like large lily pads. Rocks are also visible at the base of the American falls. The two other watercolours, neither of which appears to have been engraved, were both executed around 1766 and show different views of the falls. One is from below and again features Native people in the foreground. The other view, *Niagara Falls from Above*, provides a strange perspective, as if the viewer were standing (along with a Native chieftain and a woman) on one of the small islands in the river above the horseshoe falls.

PLATE 46: Ignace Fougeron (fl. *ca* 1760–1768), after Thomas Davies (*ca* 1737–1812), ***An East View of the Great Cataract of Niagara***, *ca* 1768. Engraving. Inscription: 'To his Excellency Lieutt: Genl: Sir Jeffrey Amherst, Knight of the Most Honourable Order of Bath, &c. &c. &c. These six views are most humbly Inscribed, by his Excellency's most devoted Servt: Thos: Davies. An East View of the Great Cataract of Niagara. Perpendr: Heigth of the Fall 162 Feet, Breadth about a Mile & a Quarter. Drawn on the Spot by Thos: Davies Capt: Lieutt: in the Royal Regt: of Artillery; J. Fougeron sculp.' From the portfolio *Six Views of Waterfalls* (London, *ca* 1768).

Soaring over the waters are two bald eagles; a third is perched on the limb of a tree. The eagles are surprisingly well executed, presaging Davies's skill as a painter of animals and birds in his later years (plate 47).

Davies's interest in natural history was obvious even in his earliest watercolour sketches, such as that of the siege of Louisbourg in 1758 (plate 2 in Hubbard), where the foreground shows a large variety of plants. It is no surprise to learn that three of the engravings in Davies's waterfall series were executed by Peter Mazell, Pennant's engraver, and Mazell may have first introduced Davies to Pennant, as well as to John Ellis, Joseph Banks, and Dr John Latham (1740–1837). From 1767 on, when he returned to England after his second

PLATE 47: Thomas Davies (*ca* 1737–1812), ***Niagara Falls from Above***, *ca* 1766. Watercolour on paper. This watercolour was not engraved. © Collection of the New-York Historical Society.

North American tour of duty, Davies became one of the circle of correspondents and informants of the European naturalists. In 1770 he prepared a paper on preserving birds that was sent to John Ellis and read at the Royal Society, then published in *Philosophical Transactions*. At some point in the next few years, he must have begun to assemble his private museum, which, according to John Latham, contained 'many scarce specimens, especially from North America, which he has been at pains to collect and arrange himself.'[18] Plants appear in many of his paintings, and Davies seemed to have been inordinately fond of plantains. The Native people referred to the plantain as 'white man's foot,' since it is an alien and sprang up wherever Europeans had settled. Davies also pictured sunflowers in a 1760 watercolour of Fort La Galette (plate 12 in Hubbard), wild cucumber in a view of the falls of the 'Seneca River' (plate 16 in Hubbard), grapevines in a 1762 watercolour of Montreal (plate 18 in Hubbard), and prickly pear in a 1778 view of New York, Long Island (plate 31 in Hubbard). He had also begun to paint exclusively natural-history subjects, and between 1771 and 1774, Davies exhibited five flower pieces at the Royal Academy, two of which showed plants 'collected in America.' In 1781, he was elected a member of the Royal Society. In 1784, he wrote to Joseph Banks from his post in Gibraltar concerning the amount of rainfall. At some point before the second edition of Pennant's *Arctic Zoology* was published in 1792, Davies had shown Pennant a 'Horned Owl from Quebec,' and Pennant included its description as an addendum.[19] He had also begun to paint his specimens of birds, and the Earl of Derby's collection boasted a portfolio of 126 paintings of birds with names and localities, executed by Davies between 1763 and 1812. A number of his paintings also ended up in the collections of Dr Latham. He published several of his bird paintings in *Transactions of the Linnaean Society* in the 1790s, and also published 'An Account of the Jumping Mouse of Canada' in the same journal in 1798. He had observed this relatively rare mouse (Meadow jumping mouse, *Zapus hudsonius*, called by Davies *Dipus canadensis*) during his last stay in Canada, and had painted it from life. An engraving accompanies the article (plate 48).

Davies's style has been described as that of an eighteenth-century Douanier Rousseau,[20] but it cannot be categorized as either naïve or primitive. Rather, the way in which Davies depicts his views and the information they contain is allied not just to the topographic style of the military artists, but to the approach of accurate observation which characterized the work of natural-history illustrators like Edwards. Davies's watercolours of Niagara Falls betray the same conventions that colour Edwards's birds, and these conventions are to a certain extent also those which characterized the English school of watercolour begun by Paul Sandby. Sandby, with whom Davies was likely acquainted and whose works define the new approach to depiction of landscape, was interested, according to his son, in 'giving his drawings the appearance of nature as seen in a camera obscura.' Gainsborough acknowledged Sandby as 'the only Man of Genius' to paint '*real views* from Nature in this country.'[21] Just as Edwards and Paillou and Griffith worked 'from Nature' when they produced accurate drawings of specimens, so did Sandby and the watercolourists with whom he was associated strive to paint accurate views of the countryside.[22] Davies's watercolours show the artist himself seated on the rocks by the river, pencil in hand, sketching the scene before him (plate 48 in Hubbard).[23] Views like those produced by Davies are not fanciful studio illustrations of idealized scenery, but renderings of nature

PLATE 48: After Thomas Davies (*ca* 1737–1812), *Dipus canadensis* (The Jumping Mouse), 1798. Hand-coloured engraving. From 'An Account of the Jumping Mouse of Canada. Dipus Canadensis,' *Transactions of the Linnaean Society*) 4 (London, 1798). Davies's paper on the jumping mouse (*Zapus hudsonius*) was read in 1797 when Davies was a major general and fellow of both the Royal Society and the Linnaean Society. Davies recounts that he observed, and thus possibly painted, the animal during his last tour of duty in Quebec (1786–90). He may well have drawn the animal in its torpid state from a preserved specimen, since he notes that he received the torpid mouse enclosed in a ball of clay from a workman in May 1787. He kept it 'until I found it began to smell: I then stuffed it, and preserved it in its torpid position.'

observed, the landscape laid down in such a way that it is both recognizable and informative.

This was in essence the goal of both scientific and military draughtsmanship – the use of the image as a medium for information. Thomas Falconer (1738–1792), friend to both Banks and Pennant, wrote to Banks that, if he planned to make any excursions on to foreign soil, he would do well to take with him a 'designer,' not to paint 'Towns or Churches,' but to illustrate the landscape: 'and if your designer would stain his drawing, it would point out the colour of the Soil and verdure, with the nature of the Rocks, and would enable us here to have a full idea of the Country, which no description possibly can.'[24] Falconer later praised Banks for his use of artists during his expedition to the Hebrides: 'Your precision of measures, and the advantage of able artists, are a great point, for when we judge by description we form an opinion through the medium of another man's understanding, who generally compares it with something else he has seen ... What an assistance is it then to truth to have the objects delineated by one common measure which speaks universally to all mankind.'[25] Again, art is the universal language, or common measure, within which science, or at least the scientific perspective, can be expressed. The illustration accords with a precise textual description, which is not a subjective account of the viewer's feelings but an accurate account of observations made on the spot. The inclusion of 'measures,' noted by Falconer, is significant, for it implies an ability to make objective descriptions of the scene; thus, Kalm spares no pains to measure the falls, and Davies includes the perpendicular height in the text accompanying his engravings, just as Paillou ensured that the exact measurements of the specimen were included on his watercolours. As Derham noted, those who ransacked the globe and its productions should be prepared to 'examine them with our gauges ...'

Topographic artists, like natural-history painters, were not working in the tradition of landscape advocated by Sir Joshua Reynolds, painting an idealized scene; rather, they were engaged in producing accurate and recognizable landmarks, for other observers whose interest in 'truth' outweighed their desire for beauty. Davies is an early practitioner of a way of rendering landscape scientifically, of producing, not beautiful scenes, but formal and accurate views which stood to textual description in the same relation that Edwards's truthful rendering of a bird stood to the careful and formalized textual description of the specimen. Davies paints Niagara from three very different perspectives, both in an attempt to come to terms with the immensity of the phenomenon, but also in the manner of the illustrator of the specimen, who presents a number of different views for the reader, so that no information is lost. There is as well a quality in Davies's paintings of falls that make them appear curiously stilled. Painting moving water is difficult, but Davies's formal rendering of the currents is akin to the manner in which he depicts rock formations, showing the strata with great precision. Bernard Smith notes a similar approach to impressive and unusual scenery in Mazell's engravings of John Frederick Miller's (fl. 1785) drawings of basalt formations in Staffa, Scotland: 'It is to be noted ... how Miller, the draughtsman and Mazell, a specialist in engraving subjects of natural history, have rendered the shape and articulation of the basaltic pillars with the same care and precision that they were accustomed to expend upon the illustration of a new species of plant or animal. The landscape is, in one sense, a scientific diagram ...'[26] The symmetrical arcs of the rainbow, the curved lines of water, and the strictly parallel strata of the rocks in Davies's paintings of falls recall the engraved feathers of Catesby's birds and the precise contours of engraved shells. Davies's landscape is not a drawing of a cascade with its fugitive light, bewildering movement, and overpowering noise; it is a drawing of a specimen, a waterfall catalogued and described in an inventory of the falls of Canada.

Animated Nature

When Thomas Pennant wrote the *Arctic Zoology*, he relied, as we have noted, on the collections and field observations of a number of correspondents resident in North America. In his work, Pennant acknowledges the help of the celebrated Dr Alexander Garden, as well as material from Samuel Hearne, Andrew Graham, and Thomas Hutchins, and we have noted his use of Davies's specimens. Pennant also refers to records and specimens from the Cook voyages, and to material he received from Anna Blackburne (1740–1783), correspondent of Linnaeus, Banks, and others, whose brother Ashton was settled in the New York area and provided her with specimens for exchange with European collectors. As well, Pennant cites works on North America by authors such as Cadwallader Colden, Mark Catesby, Pehr Kalm, Henry Ellis, Bartram, de Charlevoix, de Lahontan, DesBarres, and Josselyn. The information that came from this new group of observers was to a certain extent responsible for changing the emphasis in the writing of natural history. Often lacking access to extensive libraries, herbaria, or collections of specimens, they nevertheless had the advantage of first-hand observation, as well as Native informants. As a result, there is after 1770, at least in the natural-history literature published in English about North America, a shift from the simple documentation of the physical appearance of the specimen to an interest that goes beyond taxonomy and provides a true natural history of a new species. When Edwards published his 'Gleanings' of natural history, illustrating birds whose provenance was unclear, his purpose was the simple documentation of their appearance in a printed format, to aid other naturalists in identification. Pennant was still concerned with identification, but he could access a much broader base of material. He included in the birds section of *Arctic Zoology* information on name and synonyms and any text references, plus description (character), including varieties, then place or habitat, song, the appearance of nests, and food – what Pennant called 'manners.' This information could not be derived from the perusal of the specimen alone, nor from research into European literature. These were new data, derived from the accurate inspection of new species by observers on the spot. Pennant's work is also one of the first published examples of a zoogeographical approach to understanding natural history. The difficulties of acquiring and preserving birds, animals, fish, and so on, had made assembling a representative selection of a region's animal life extremely difficult. By the combination of literature review, reports from resident observers, and examination of both living animals and documented specimens, Pennant was able to achieve a remarkable picture of the animal life in the polar and subarctic regions. His work includes a table showing distribution of quadrupeds in the Old World and the New, and acknowledges that more information is wanting on migration of birds and mammals. A zoogeographic approach demanded new methods of collection, and Pennant complained, for example, that, on Cook's third circumnavigation, 'it was a great misfortune, in this voyage, that the fishes were promiscuously flung into one common case, so that it is impossible to ascertain the species belonging to each country.'[27]

More than the works of any cabinet naturalist, then, Pennant's *Arctic Zoology* depended on the observations and collections of the amateur field naturalists, to whom Pennant pays tribute as the 'gentlemen or writers who have paid no small attention to their [the specimens'] manners.'[28] Pennant's field informants worked, however, under some disadvantages. Few of them had received formal training in natural history, but many were eager, including Thomas Hutchins, who, acknowledging his own lack of training, yet happily presented his remarks to the Company as an accompaniment to a collection prepared by Andrew Graham and himself: 'The following sheets were not wrote by one who is skilled in zoological affairs, but by a young person seeking after knowledge and improvement, who would think himself extremely happy to be of service to the learned, and is proud of every opportunity of demonstrating his gratitude for favours received from the Hudson's Bay Company.'[29] Humphrey Marten, who arrived in Hudson Bay in 1750 and remained with the Company until his retirement in 1786, sent a collection of birds to the Royal Society in 1772, with an explanation of his 'willing mind and weak ability.' His comments on the life of the collector deserve to be repeated at some length, since they so accurately reflect some of the difficulties under which these willing servants of the Company and natural history laboured:

> I hope those Gentlemen that inspect the aforewritten history of birds will observe, that I do not absolutely declare all I have set down to be truth ... I was obliged to have the best Indian intelligence I could get. I trusted not to assertions of any single person, let his age or experience be what it would, therefore hope that I am near the truth, if not quite so. I could have wished that when I received orders from my masters to make a collection of birds etc. that the Naturalists' Journal, as also the British Zoology, had been sent to me, for which I could have paid with thanks; fine seed shot, birdlime, glass bottles with ingredients for making the preparing liquor, would have enabled me to have given more satisfaction to the gentlemen concerned as well as myself than it is possible for me now to do. Wire, and beads of all sizes and colours that resemble the eyes, I presume should have been sent: as not a soul I believe in Hudson's Bay knows anything of painting either in oil or water colours.[30]

Some of the observers were more able than others, and Samuel Hearne in particular stands out as being very able indeed. When the Comte de la Pérouse took Churchill for the French in 1782, he captured not only Hearne, but his journals. So impressed was the Count with the journals that he returned them to Hearne on condition that he publish them. Hearne's journals were published in 1795 after his death as *A Journey from Prince of Wales's Fort in Hudson's Bay to the Northern Ocean 1769, 1770, 1771, 1772*. Included in the book are a number of corrections to Pennant's work. Pennant had obviously had access to Hearne's manuscript notes, for he frequently cites his observations and the descriptions of 'his elegant pen.'[31] Pennant did not, however, check everything against Hearne's first-hand knowledge, since, according to Hearne, a number of false observations have 'crept into Mr. Pennant['s] Arctic Zoology ... that elegant work ...'[32] (One of the hoariest old chestnuts repudiated by Hearne was the oft-repeated and illustrated tale of beavers using their tails in the manner of masons with trowels. Hearne writes that he 'cannot refrain from smiling, when I read the accounts of different Authors who have written on the oeconomy of those animals, as there seems to be a contest between them, who shall most exceed in fiction.')[33] Pennant was aware of some of the shortcomings of his methods, and apologized in his advertisement that his reliance on amateurs had of necessity made his work less perfect. He regretted that a brilliant and trained naturalist like Willughby had failed to achieve his planned voyage to the New World: 'What he [Willughby] would have performed, from an actual inspection in the Native country of the several subjects under consideration, I must content myself to do, in a less perfect manner, from preserved specimens transmitted to me ...'[34] Rheinhold Forster (1727–1798), who prepared some of Graham's specimens for publication in *Philosophical Transactions*, also noted some of the problems in working with the untrained amateur:

> It is however presumed, that though Mr. Graham has shown himself a careful observer, and an indefatigable collector, yet, not being a naturalist, he could not enter into any minute examination about the species to which each goose belongs, nor from mere recollection know, that his grey goose was actually to be met with in England. A natural historian, by examination, often finds material differences which would escape a person unacquainted with natural history. The wish, therefore, of seeing specimens of these species of geese, must occur to every lover of science.[35]

Despite the new information on animal behaviour and geographic range which Pennant was able to include as a result of his correspondents' efforts, as Marten noted, the lack of trained artists seriously precluded the depicting of animals and birds in their environment with any degree of accuracy. Pennant was forced to rely on the depiction of mounted specimens, though it is obvious from the marginalia by Mercatti and Giffith included in his personal copy of *Arctic Zoology* (now in the Blacker-Wood Library at McGill University) that he could have wished for more felicitous renderings of both scenery and animals. The frontispiece to the introduction by Peter Paillou does, however, make an attempt to depict the arctic wastes with their characteristic inhabitants (plate 49). Paillou, it should be remembered, had painted Pennant's hall at Downing with 'several pictures of birds and animals, attended with suitable landscapes,'[36] and it may be that one of these served as model for

PLATE 49: Peter Mazell (fl. 1761–1797), after Peter Paillou (fl. 1744–1784), Frontispiece, 1783. Engraving. From Thomas Pennant, *Introduction to the Arctic Zoology* (London, 1792). Pennant's work was one of the first efforts at zoogeography, and his frontispiece is described as a 'winter scene in *Lapland*.' The animals are, however, characteristic of the Canadian arctic, and Pennant obtained many of his specimens from his connections with the Hudson's Bay Company.

the frontispiece. The frontispiece is described as 'a winter scene in *Lapland*, with *Aurora Borealis*: the Arctic Fox, Ermine, Snowly Owl, White Grous.' But just as Pennant's book is panarctic, so the frontispiece stands for a generic arctic landscape. The engraving shows a rocky crag, and perched on it is a snowy owl, talons clenched firmly around the body of the 'grous' (ptarmigan). Below the owl squirms a pale arctic fox in a contorted position beneath a rock. Another owl hunches to the left of the fox, while at the bottom right an ermine slips by the ensemble.[37] The northern lights glimmer in the background where can barely be seen a lake and a Lapp village. There is an obvious staging in the dramatic poses of the animals on their rocky hillside, reminiscent of the manner in which a variety of mounted birds were perched inside glass domes or cabinets. The ensemble lacks the casual realism of Davies's eagles soaring above the falls, and points out the importance not only of the trained observer, but of the trained artist making drawings 'on the spot.' Getting the animal 'right' by painting from a prepared specimen was one thing, but fitting the animal or animals into their habitat and depicting their relations one to another required another sort of understanding altogether, one that could not be achieved by either the cabinet naturalist or the studio artist. As with Kalm's account of Niagara, so Pennant's natural history was hampered by the limitations of the illustrations which accompanied it, the sixteen plates (fourteen of which are obviously of mounted birds) recording specimens, not 'manners.' The dissociation between text and illustration could be remedied only by the kind of artistic and scientific collaboration that had made James Cook's (1728–1778) circumnavigations milestones in the development of natural history.[38]

Joseph Banks and the Cook Expeditions

Cook's first circumnavigation was promoted by the Royal Society and one of its purposes was to observe the transit of Venus in 1769. A party of scientists and artists organized by Joseph Banks (elected a member of the Society just before his departure for Newfoundland) was included at the request of the Society. It is possible that Cook and Banks had met in 1766 in Newfoundland, when Cook had been charged with the Newfoundland coastal survey, and it was Banks's voyage to the island and Labrador that prepared him for the much more extensive exploration two years later on the *Endeavour*. For his

Newfoundland trip with his friend Constantine John Phipps (1744–1792), he took with him a small library that included works of Linnaeus, Edwards, and Catesby, as well as assorted collecting equipment, such as nets and trawls, butterfly nets, and plant presses, as well as a keg of spirits to preserve animals and birds. Banks must have realized from this voyage the disadvantages to accurate rendering which resulted when no artist was available. Some of his plant specimens were not properly dried, owing to haste and illness; others were destroyed in heavy seas. Although Ehret's watercolours on vellum of the Newfoundland plants are remarkable in their accuracy, considering that they were prepared from dried specimens, mistakes did occur when painting from herbarium material. The *Dryas integrifolia*, for example, is painted with yellow flowers. The flowers are actually white, having turned yellow when dried. Sydney Parkinson (1745–1771) was responsible for painting the bird specimens from the trip, and succeeded so well in rendering the preserved specimens that Banks employed him as the natural-history artist on his round-the-world expedition, on which Parkinson died. Parkinson was one of only two artists on the voyage, the other being Alexander Buchan, a topographical artist. Buchan, too, died shortly after the expedition arrived at Tahiti, and Banks wrote in his journal that 'his loss to me is irretrievable, my airy dreams of entertaining my friends in England with the scenes I am to see here have vanished. No account of the figures and dresses of the Natives can be satisfactory unless illustrated by figures ...'[39] Luckily, Parkinson was able to fill the void left by Buchan's death, and both Daniel Solander and Herman Spöring, the two naturalists, also drew scenes and specimens.

Banks's enterprise in ensuring a scientific and, above all, illustrative record of the circumnavigation was not lost on the navy. As early as mid-century, Richard Walter, in a popular edition of Anson's *Voyage*, had pointed out how much had been lost by not ensuring better visual records of important voyages:

> I cannot ... but lament, how very imperfect many of our accounts of distant countries are rendered by the relators being unskilled in drawing, and in the general principles of surveying; ... Had more of our travellers been initiated in these acquirements ... we should by this time have seen the geography of the globe much correcter, than we now find it; the dangers of navigation would have been considerably lessened, and the manners, arts and produce

> of foreign countries would have been much better known to us, than they are. Indeed, when I consider, the strong incitements that all travellers have to acquire some part at least of these qualifications, especially drawing; when I consider how much it would facilitate their observations, assist and strengthen their memories, and of how tedious, and often unintelligible, a load of description it would rid them, I cannot but wonder that any person, that intends to visit distant countries, with a view of informing either himself or others, should be unfurnished with so useful a piece of skill.[40]

Drawing had been included in the curriculum of the Portsmouth Naval Academy in 1733, and as a result of the publication of Anson's *Voyages*, commanders were given orders to ensure that 'officers skilled in draughtsmanship' would take, wherever possible, coastal profiles and plans of harbours, fortifications, and anchorages (plate 50).[41] The inclusion of artists and naturalists, or at least of naval officers trained in science and art, became customary on the major exploratory voyages of not only the British, but also the French, Spanish, and Russians (plate 51). The paintings and sketches provided by these artists were made not only to accompany the specimen collections, but with an eye to subsequent publication. Bernard Smith chronicles the depiction of Australia and its natural productions by naval artists, but in North America, with the exception of the West Coast, most of the illustrations to travel accounts were the result of the activities of military artists such as Davies. Even the coastal survey of the East Coast relied on the talents of an infantry officer trained at Woolwich. Joseph Frederick DesBarres (1722–1824) was drafted into a naval survey of the coasts of Nova Scotia, which resulted in the *Atlantic Neptune* (1780 edition), a survey still in use (as were many of Cook's charts) until well into the nineteenth century. DesBarres included a series of views in the publication, such as the windswept image of the surveying camp on Sable Island, complete with wild horses and monstrous sand hills rising like sugar loaves over the scene (plate 52).[42] The exception to the predominance of soldier-artists was in the depiction of the arctic regions, where the navy had declared its interest. Almost forty years after Pennant had first published his record of the manners of the beasts and birds of the region, and had wished for the 'potent emanations of a Linnaeus's to inspire the efforts of an 'American philosopher,'[43] trained artists and naturalists at last provided the figurative counterpart to the detailed text descriptions of Graham and Hutchins and the 'elegant pen' of Hearne.

PLATE 50: ***Views of Parts of the Coast of North West America, 1798.*** Engraving. Inscription. 'VIEWS of PARTS of the COAST of NORTH WEST AMERICA T. Hoddington del^t The Westernmost of SCOTS ISLANDS ... J. Sykes del^t CAPE SCOT ... J. Sykes WOODY POINT ... H. Humphrys del^t The entrance of NOOTKA SOUND ... H. Humphrys The entrance of COLUMBIA RIVER CAPE DISAPOINTMENT ... H. Humphrys PUNTO BARRO DE ARENA W. Alexander del^t from Sketches made on the Spot London. Published May 1st by R. Edwards New Bond Street J. Edwards Pall Mall and G. Robinson Paternoster Row Engraved by B. T. Pouncy.' From George Vancouver (1757–1798), *A Voyage of Discovery to the North Pacific Ocean and round the World ... Atlas of Plates* (London, 1798).

PLATE 51: After Aleksandr Filippovich Postels (1801–1871), Frontispiece: *Algarum vegetatio*, 1840. Lithograph. From Aleksandr Filippovich Postels and F. Ruprecht, *Illustrationes Algarum in itinere circa orbem jussu imperatoris Nicolai I* (St Petersburg, 1840). One of the most brilliant nineteenth-century depictions of the natural historian. Postels was the expedition artist and also made a collection of algae. His view of the specimens below water and of the scientist and his assistant above, with the scenery of the Northwest Coast as a backdrop, is startling. Postels's large book is written in Latin and provides full-scale lithographs of the algae collected.

PLATE 52: After Joseph F.W. DesBarres (1722–1824), ***A View from the Camp at the East End of the Naked Hills ... on the Isle of Sable, Nova Scotia***, *ca* 1779. Aquatint. From Joseph F.W. DesBarres, *The Atlantic Neptune* (1780).

The First Franklin Overland Expedition, 1819–1822

By the end of the eighteenth century, coastal surveys of the east and west coasts had been completed by Cook and Vancouver, but the Arctic coasts were still uncharted, and the Northwest Passage, so ardently sought since the age of Elizabeth, remained a mystery. For most Europeans the very appearance of Hudson Bay and the Arctic islands was equally mysterious. Mention should be made again of the 1768 published engraving *Inhabitants of North America near Hudsons Bay ...* (plate 34) that was based, albeit loosely, on John White's engraving of two centuries earlier (see chapter 4). A number of expeditions in search of the Northwest Passage had recorded the icy landscapes, and Henry Ellis's account of an expedition in 1746–7 in the *Dobbs Galley* and *California*[44] included some illustrations of the landscape as well as of the bird and

PLATE 53: ***A Winter View in the Athapuscow Lake***, *ca* 1796. Engraving. From Samuel Hearne (1745–1792), *A journey from Prince of Wales's Fort in Hudson's Bay to the Northern Ocean* ... (Dublin, 1796), 248.

animal life. The 'cuts' of the landscape were at best crude schematic depictions of points in the narrative, while most of the illustrations of animals were culled from George Edwards, to whom Ellis gives appropriate credit. Samuel Hearne and Alexander Mackenzie had both made major overland expeditions in the late eighteenth century, and Hearne's observations had provided at least some knowledge of the Arctic coastline around the mouth of the Coppermine River. As has been noted, Hearne was a careful and meticulous observer, but his book contains only a single illustration, which may have been engraved after Hearne's own sketch, or might have been prepared by the engraver from textual description alone. Certainly the engraving *Winter View in the Athapuscow Lake* (plate 53) captures something of the nature of the northern lands, though, as Maclaren notes, the trees are far too symmetrical and disproportionate to be a true representation of the scene.[45]

In the second decade of the nineteenth century, the Royal Navy, under-employed after the end of the Napoleonic Wars, turned its attention to the Canadian Arctic. The success of the Cook expeditions had not been forgotten, and participants in many of the Arctic voyages were selected as much for their artistic and scientific abilities as for their seamanship. In 1818, John Ross (1777–1856) was sent to resurvey Baffin Island, and a year later another expedition under Parry and Liddon sought the Passage by sea, but, at the same time, in a departure from custom, the Navy embarked on an overland expedition using naval officers. Led by Lieutenant John Franklin (1786–1847), the surveying party included two midshipmen, George Back (1796–1878) and Robert Hood (1796–1821), and a naval surgeon, Dr John Richardson (1787–1865). Richardson's assigned duties were to 'collect and preserve specimens of minerals, plants and animals,'[46] while Back and Hood were charged with making accurate observations and providing the illustrative records of the expedition. Both Back and Hood were employed making charts of the expedition's progress, recording temperatures, and observing the weather, but it is obvious from their journals and their original sketchbooks and watercolours that Hood was the more expert in depicting natural history, and Back the better at rendering landscape views. Back's two sketchbooks from the expedition include fifty-seven watercolours and pen-and-ink drawings, and a number of these form the basis for the illustrations by Edward Finden to Franklin's account of the expedition. Back's sketchbooks also included maps with notations of topographical and geological features, as well as compass directions, temperature records (many taken from Hood), and distance travelled. Hood's sketchbooks appear to be no longer extant, perhaps having been lost with Franklin's writing desk and papers in the crossing of a river on the disastrous return trip from the shores of the Arctic Sea.[47] A number of Hood's full-size watercolours survive, however, having likely been included with a packet of materials sent to England in March or July 1821.[48] Hood's watercolours were also used by the printers for the Franklin expedition publication.

The great naval voyages of the eighteenth and early nineteenth centuries made an enormous impact on natural history, due in part to the quantity of specimens collected and illustrated, and the number of observations and records made. This success was in many ways the result of their mode of transport. A naval ship was like a floating laboratory, well equipped to house the specimen kegs and plant presses, and carrying adequate stores of paper both for herbarium sheets and for watercolours. It also provided shelter and space for the artists and, during the long passages from landing to landing, an opportunity to produce finished sketches. Circumstances were very different during an overland expedition, particularly one to the far north, and artists such as Hood and Back were faced with great difficulties both in keeping their journals and in making their records. Back's surviving sketchbooks from the 1819–22 expedition show something of the way he and Hood must have worked. Back's sketchbooks are small, the largest being 11.43 by 19.05 centimetres. One is a journal, but the other is a formal sketchbook, with glazed paper separating the heavier leaves. Back used both for maps, sketches, watercolours, and notations. He worked in the style of most watercolour artists, first making a pencil sketch, then laying either washes or full-body colour over the pencil lines. Maclaren notes that Back was no doubt aware of the engraver's requirements for defined separations between light and dark areas, and prepared his works accordingly, in hopes that they would be included in the official printed record.[49] (It is obvious from the style of their journals that Hood, Back, and Richardson all looked forward to publication, since the expedition was supported by the British government.) Pencil sketches were often made on the spot, as can be seen in the ink drawings of a buffalo showing the head, hoof, and detail of hindquarters, which may be the record of a buffalo shot on 6 February 1820, when Back noted: 'I took a sketch of it directly.'[50] Back also notes that in some cases he made sketches outdoors, such as the one of Fort Carleton made on 7 February of the same year, and he depicts himself, in much the same way as Davies, sketching by the side of the Upper Falls of Wilberforce Falls (plate 54) and by a rapid in the Coppermine (2 July 1821). On-the-spot records were not, however, always possible, for reasons both of the extreme cold and of the incessant swarms of insects. On 29 December 1820, Back records in his journal the crossing of a series of portages. Normally responsive to the picturesque views presented by waterfalls (as Davies and a horde of other watercolourists of the period were), Back notes that 'when a person can just prevent himself from freezing – he has no great relish even for the finest views.'[51] If it was not the cold, it was the insects. On 10 July 1821, on the shores of the Coppermine River, Back recorded that he returned with the others to their tents 'amongst myriads of mosquito's, which in spite of veils – gloves – or handkerchiefs stung us in a most unmerciful manner – and quite prevented any attempt to sketch.'[52] Maclaren observes that Back's sketchbooks do not show any signs of insect kills, and notes that Back may have mixed in his colours a bitter substance called coloquintida, sometimes used by watercolourists to prevent insects from landing on drawings.[53]

Sketches were finished in larger format during the long stretches when the expedition was forced to remain idle, waiting for supplies, or for the lakes to freeze for winter travelling, or for the rivers to thaw for travel by canoe. Back's sketchbooks bear evidence of his forethought for the completion of finished watercolours. The colours are often included on the glazed paper. Beside a sketch of White Falls, he notes the following information: 'Rocks grey with light green. Tint of Red ochre and Gamboge. Rocks to the right, a hint of Lake. Fresh Bright tints. Much foam and spray about the nature further to the right.' On another drawing executed on 13 February 1821 at the Petite Portage on the Slave River, he notes the colour instructions with a comment on the absent foreground – 'Dogs and train passing.' Similarly in a sketch of Portage La Loche, he writes, 'the foreground omitted here, but it is drawn in the picture sent home. It forms a part of the hill road by which you descend.' A pen-and-ink outline is included in the corner.[54] Back notes that they passed a quiet early spring at Cumberland House – 'Nothing particular occurred till May when the lake began to open' – but their time 'was fully occupied in bringing up the Chart Drawings &c.'[55]

It can be assumed that a number of Hood's initial sketches were also completed in larger format during the fall and winter of 1819–20, a period of 'full occupation,' when the expedition waited out the winter at Cumberland House. Even when under shelter, Arctic artists had problems. Hood noted that, despite keeping 'the chimneys in a constant blaze,' their 'pens and brushes were frozen to the paper.'[56] Hood also described other problems encountered by the natural-history artist in the field. Towards the end of March 1820, Hood, 'desirous of obtaining a drawing of a moose deer,' sets off with some Native hunters who, on 2 April, kill a moose. Only by offering the Native people their own provisions do Hood and his companions manage to 'suspend the work of destruction until the next day.' The Native hunters had already removed the entrails and the foetal moose, and Hood prepared to sketch the carcass. His account of the episode is worth noting: 'I had scarcely secured myself by a lodge of branches from the snow, and placed the moose in

PLATE 54: Sir George Back (1796–1878), Upper Falls of the Wilberforce Falls, Northwest Territories, 27 August 1821. Watercolour. From George Back, Sketchbook II, 1821–2.

a position for my sketch, when we were stormed by a troop of women and children, with their sledges and dogs. We obtained another short respite from the Indians, but our blows could not drive, nor their caresses entice, the hungry dogs from the tempting feast before them.'[57]

It is no wonder that the resulting watercolour from the sketch made under such conditions presents a very stiff-legged and peculiarly proportioned doe moose munching on a tree against a snowy countryside. Hood's watercolour is supplemented by his journal entry, which gives measurements of the moose, and its weight, colour, and habits as learned from the Native people. Hood notes that, 'as the American moose deer is an animal of which I have not seen any perfect account, I shall add to what is already known, the information procured on this occasion.'[58] During the stay at Cumberland House, Hood also figured a wolverene, lynx, fisher, buffalo (of which Back also prepared a drawing), otter, and ermine, as well as composite sketches of birds. It would appear that Hood sketched the birds individually from dead specimens, then repainted them onto a background of trees or water. In the case of a series of winter birds, Hood has actually cut out the individual sketches and pasted them on a grey sheet. While the colouration is adequate, it is obvious from the attitudes in which the birds are posed that Hood had not thought of Audubon's trick of wiring the dead specimens into lifelike postures. The birds are stiff and unnatural, the wings spread too wide, the necks stretched too far.[59] At Fort Enterprise in 1821, Hood painted a white wolf and a series of fish. Hood also turned his hand to landscape, and completed a number of views of falls and lakes. Before his death, Hood managed to finish twenty-eight paintings. Back was more prolific, despite the fact that his winter journey of more than 1,100 miles on snowshoes in 1820–1 prevented his completing more drawings.

Back and Hood approached their assignments from two very different perspectives but in a complementary manner. Both were trained in the topographical tradition of naval documentation. Back likely learned his skills from senior officers during his time as a prisoner of war in France, while Hood may have learned how to record landscape on board ship. Both are confronting an entirely new kind of landscape and are required to depict it faithfully, and Back at least expresses confidence in his journal entries that accurate drawing is the equivalent of, if not superior to, textual description: 'The description of it will be better explained in a drawing than in any other manner ...'[60] Bernard Smith, in his exposition on the art of Cook's voyages, *European Vision and the South Pacific*, underlines the importance of topographically trained artists in creating the depiction of the 'typical' landscape, the view which laid out the 'character' of the country, in the same manner that a properly executed drawing of a flower could show its character at a glance. Smith defines the typical landscape as 'a form of landscape the component parts of which were carefully selected in order to express the essential qualities of a particular kind of geographical environment.'[61] The typical landscape is the antithesis of the idealized landscape, and the 'component parts' are selected as much on the basis of their explanatory power as for their aesthetic or picturesque appearance. Smith relates this view of nature with its emphasis on particularity to the ideas on 'geographical environment' espoused by Alexander von Humboldt (1769–1859) in a series of works published in the early 1800s. Whether Hood or Back, or even Richardson had access to these works is uncertain, but the popularity of Humboldt's *Personal Narrative* and subsequent works was evidence of the changing perspective on the study of natural history that was apparent even in *Arctic Zoology*.[62] Humboldt's works were concerned with physical geography, with the description of characteristic climatic zones. Smith quotes from the English translation by Mrs Sabine of Humboldt's *Aspects of Nature*, which was first published in 1808: 'The azure of the sky, the lights and shadows, the haze resting on the distance, the forms of animals, the succulency of plants and herbage, the brightness of the foliage, the outline of the mountains, are all elements which determine the total impression characteristic of each district or region.'[63] In the paintings of both men is reflected, then, this new approach to landscape which Smith sees as derived by necessity from the confrontation between the European-trained artists and the unfamiliarity of exotic terrains. Next to the tropics, the Arctic was probably the landscape most likely to challenge the conventions of contemporary artistic depiction, and both Hood and Back responded from the background of their training and their interests to depict the northern landscape and its natural productions and create a 'total impression' of a new landscape.

Hood's portraits of both animals and Native people reveal an awkwardness, but his interest in the habits of the Native population and the indigenous wildlife was genuine. He began a section of his journal on 'Animals' but managed to complete only an entry on the buffalo, noting that 'other occupations have frustrated the design at present.'[64] In his observations on the mosquito, which he compares to the mosquitoes of Africa and Europe, he shows as much interest in describing their natural history as Back had shown in describing their torments.[65] Hood showed a sensitivity to nature that is quite profound, and his description of an arctic spring is a revelation:

> By the 21st [May], the elevated grounds were perfectly dry, and teeming with the offspring of the season. When the snow melted, the earth was covered with the fallen leaves of the last year, and already it was green with the strawberry plant, and the bursting buds of the gooseberry, raspberry and rose bushes, soon variegated by the rose and the blossoms of the chokecherry. The gifts of nature are disregarded and undervalued till they are withdrawn, and in the hideous regions of the Arctic zone, she would make a convert of him for whom the gardens of Europe had no charms or the mild beauties of a southern clime had bloomed in vain.[66]

It is this sensitivity to the natural world that infuses Hood's depictions of the countryside and distinguishes them fundamentally from Back's. Hood's view of Marten Lake is perhaps the most visionary view of the Canadian north painted throughout the first half of the nineteenth century (plate 55). It is both the rendition of a moment in the narrative of the expedition and an incredibly powerful descriptive statement about the nature of the Arctic lands. Almost half the large watercolour is sky. Light streams through the clouds in separate beams onto the still surface of the lake. To the right, on a rocky hill, a small group of caribou stand. Below them, gliding silent through the reflection of the hill, is a voyageur canoe. One of the expedition members stands in the canoe, ready to fire on a caribou swimming for the shore. The movement has alarmed a flock of snow geese, who rise skywards in two parallel rows into the beams of light. In the foreground, two Native hunters crouch behind cover, eyeing a wary herd of caribou on the left. The animals are magnificent, with full spread of antlers. The watercolour has faded from exposure to light, but it remains clear, pure, and mystical, filled with acute observation and immense appreciation for the landscape. But it is Hood's inclusion of the animal and bird life of Marten Lake that makes the watercolour remarkable. In all his depiction of animals, Hood uses the landscape, not as background, but as habitat. The buffalo kicks up the dirt of the prairie, the

PLATE 55: Robert Hood (1796–1821), ***A Canoe on the Northern Land Expedition Chasing Reindeer in Little Marten Lake, Northwest Territories***, 1820. Watercolour on paper. Inscription: 'A canoe of the north land expedition in Marten lake, chasing Rein deer, Lt. 64 42′. N. Long 112 5′ W. September – 1820 An Indian in the fore ground, creeping towards the deer. Robert Hood. Fort Enterprize. April 1821.' Hood's luminous watercolour is remarkable for his vision of the north, filled with light and life. Hood describes the snow geese which feature so prominently in this view: 'The gale brought with it, from their northern haunts, long flights of [snow] geese, stretching like white clouds from the northern to the southern horizon, and mingling their ceaseless screams with the uproar of the wind among the hills.'

otter slips into a frozen stream, the lynx attacks a hare near the forest edge. To Hood the Arctic is an inhabited world, and his watercolours presage the ecological depiction of birds and other species that are the hallmark of the work of Audubon, and later of John Gould.

While Richardson's journal is a relatively terse account, with detailed notes on plants and animals encountered and extensive recording of geological information, Hood's journal entries also include purely descriptive passages which reveal not just the naturalist's, but the artist's, eye,

> We were prepared to expect an extensive prospect, but the magnificent scene before us was so superior to what the nature of the country had promised, that it banished even our sense of suffering from the mosquitoes which hovered in clouds about our heads. Two parallel chains of hills extended towards the setting sun, their various projecting outlines exhibiting the several gradations of distance, and the opposite bases closing at the horizon. On the nearest eminence, the objects were clearly defined by their dark shadows; the yellow rays blended their softening hues with brilliant green on the next, and beyond it, all distinction melted into grey and purple ... Impatient as we were, and blinded with pain, we paid a tribute of admiration which this beautiful landscape is capable of exciting, unaided by the borrowed charms of a calm atmosphere glowing with the vivid tints of evening.[67]

It is interesting to compare Hood's relatively restrained if sensitive description of the view with that of Back, taken from a similar vantage point:

> Breaking through the thick foliage of pine and cypress – he stands at once on the summit of an immense precipice – and like one bewildered in some vast labyrinth – knows not where to fix his eye – The view is that of a valley some thousand feet beneath you, in length upwards of 30 miles – in breadth three or four – and in the centre a meandering river holds its course near, covered with snow – farther distant, less light – decreasing in brightness – till it becomes insensibly lost in the deep blue mist of the distant perspective – it is bounded on each side by immense hills – the fragments of gigantic mountains – irregularly broken – like to the confusion of an earthquake – undiscribably [*sic*] grand – the one side is burnished over with the luxuriant

PLATE 56: Sir George Back (1796–1878), 'Common Tops to the North,' 1822. Watercolour. From George Back, Sketchbook II, 1821–2.

> foliage of the pine and cypress – the other with the cold and sterile poplar – half lost in snow – while here and there on the summit – scattered promiscuously – dark towering forests hang half suspended ... I do not pretend to describe the beauties of this view – the pencil being a more powerful vehicle than the pen for that purpose, for the whole is apt to vanish before the minute parts can be described ...[68]

Back's description is part instruction to the artist and part romantic gloss. Maclaren notes that Back's carefully contrived watercolours (pine trees are practised in the sketchbooks [plate 56]) of typical northern landscapes sit uncomfortably with his descriptions of wild, savage country.[69] Earthquakes, cold, sterile poplars, luxuriant foliage, dark towering forests – this is the language of romanticism, a view of the Arctic as a place of mystery and beauty. This type of description was at odds both with Back's training and with his ability to depict landscape. It is, however, revealing about an aspect of Back's

PLATE 57: Sir George Back (1796–1878), Northeast View of Great Slave Lake from Fort Providence, Northwest Territories, 10 November 1820. Watercolour. From George Back, Sketchbook I, 1819–20.

personality, and his reponse to landscape. To the naturalist like Richardson, every scene has its observations and its specimens. It was obvious to Richardson, however, that Back appreciated particular kinds of scenery. In a letter to Back, Richardson writes, 'Amongst these hills you may observe some curious basins, but nowhere did I see anything worthy of your pencil. So much for the country. It is a barren subject, and deserves to be thus briefly dismissed.'[70] From his journals and his sketchbooks, it is obvious that Back responded most strongly to a distinctive type of scenery, 'diversified,' with 'singularity,' that might 'catch the eye of the traveller.'[71] Clumps of willows on the great plains, in his opinion, 'gave some relief to the continual eveness [*sic*] of the scene ...';[72] otherwise, as he noted *en route* to Fort Chipewyan from Fort Enterprise, 'the scenery is so like what we had already passed that I cannot imagine how one part is distinguished from the other.'[73] Like the picturesque traveller, Back is on the lookout for 'views,' and his choice of what he paints often reveals this bias. Falls and rapids figure largely among his sketches, and his illustrations of the Arctic sea passage in canoe show towering cliffs, stormy seas, and lowering skies. As a trained observer, however, Back does not sketch only spectacular scenery. In November 1820, he paints a panorama of Great Slave Lake, with a view to the northeast, to the southeast, and to the southwest (plate 57). Back completes the set in 1821, with a view to the northwest. He also paints two rocks covered in lichen (*Tripe de roche*, which featured so greatly in the survival of the party), as well as pencil and pen-and-ink sketches of beaver and muskrat, studies of caribou skulls and antlers, and depictions of typical plants – crowberry, blueberry, and swamp tea. Even Back's paintings of rapids and waterfalls are more than simple picturesque views. Like Hood's, they include with precision and care the typical features of the new landscape in all its particularity.

Hood's watercolour of evening on Marten Lake was a record of the typical landscape of northern Canada, complete with 'the forms of animals, the succulency of plants and herbage, the brightness of the foliage, the outline of the mountains' characteristic of the region, but it was also something more. It showed a country of incredible beauty far removed from the European scenes in which both Hood and Back had grown up. The fact that both men could depict this new landscape with sensitivity and a regard to particularity indicates a fundamental change from the understanding of the country expressed two hundred years earlier by du Creux, when he wrote that many people had tried to discourage him from writing a history of Canada, 'a subject at once so meagre and so grim ...' How could there be anything to describe, any history, in a land devoid of cities, palaces, gardens, canals, and aqueducts? What were its landmarks, its monuments?[74] The illustration of natural history, and by extension of habitat, provided Europeans with a new understanding of the meaning of place. They were able to give to du Creux's 'immensity of woods and prairies' a shape which they could define and describe, from which they could isolate a landscape of meaning. By using the conventions of natural history, painters such as Back and Hood could render the unfamiliar and the new into a readable document for the eye of the observer.

Conclusion: Drawing and Nature

I began this enquiry with a series of questions about a single image. Looking at that particular and singular image, '*The Bittern from Hudson's-Bay,*' led me to a more general examination of an entire class of images, and an attempt to comprehend the ways in which these images had been created, used, and understood by their viewers or, some might say, their readers.[1] My original questions focused on the nature of a dissonance I perceived to exist between the way in which we, as late-twentieth-century readers, saw Edwards's rendition of a bittern, and the way it was viewed by Edwards's intended audience. I was intrigued as well by the notion of accuracy in rendition which this dissonance brought to attention. What did it mean to 'draw from nature'? In my examination of images in natural history, I have tried to explore both the understanding of nature and the role of drawing in the communication of knowledge, at various points along a continuum from the sixteenth to the early nineteenth century. These points have been chosen by the selection of the images, and, as I noted in the introduction, my work has been grounded in the drawing, not the word. I have tried throughout to tease as many threads as possible out of each image, to link the visual and contextual fields of the image both forward and backward through time, tracing relationships, antecedents, persistence, and change. In this final chapter, I want to draw some tenuous conclusions about the changing understanding of nature, as revealed in the depiction of plants and animals and landscapes, and to offer some observations on the close reading of drawings and their part in the creation of knowledge.

Notions of Nature

We have long been conditioned to see nature as other, as distinct from the civilized world we construct for ourselves. In recent years, there has been a movement to encourage the adoption of a new perspective on our place in nature, to see ourselves as a species among species in a fragile environment, a perception many would contend was formed by our first images of Earth from space, a marbled blue sphere hanging luminous in the void. The perception of nature as the other and as an entity that demands attention or study is, however, like our contemporary perception of Spaceship Earth, conditioned by history and culture. That this attitude towards the natural world has not always been with us is the subject to which Keith Thomas has devoted so much careful research. He notes that 'between 1500 and 1800 ... there occurred a whole cluster of changes in the way in which men and women, at all social levels, perceived and classified the natural world around them.'[2] These changes come to light not only in what people said, but in how they pictured the natural world, and it is these changes we see revealed in the images of plants and animals and landscape documented here. I would also contend that, for many people during this period, ideas about nature were conditioned by the discovery of the New World. If anything, it forced Europeans to confront their own habitat and make comparisons between what was the same or similar and what was new and strange. That many devoted more attention to the exotic and singular should be no surprise, but what *is* surprising is the gradually awakening interest in the documentation of all species, no matter how common or familiar. What is also important to note is that this documentation relied on the ability of artists to represent something. It is a mistake to divorce the history of thought from the history of making. What we have been able to do has often conditioned what we have thought, and poor materials, rare models, ill-preserved specimens, and imperfect processes of reproduction, as well as artistic skill, have all combined to structure understanding.

In the maps drawn with such attention to detail and visual delight by the Dieppe hydrographers, the natural world is represented by isolated and emblematic images. The map-makers were, after all, map-making, attempting to present a schematic view of the outlines of the land and its landmarks. Flora and fauna were seen as marks or signs that would enable the navigators to recognize their surroundings. If that is a white bear or a dark falcon, it must be the New World. Map-making is an abstraction, and the animals and vegetation featured on these early charts are also abstracted, and, as we saw in the case of Thevet's su, even portable, serving equally in many contexts which exemplify the new and the strange. Iconic animals did not, however, satisfy the need for the explorers and authors to communicate their findings to a larger audience. They demanded renderings that would picture for their readers their own memories, or in some cases the actual plants or animals returned from the New World. The development of naturalism has been admirably treated by many authors, including Klingender and Gombrich, but what has been important in this analysis is the use made by naturalists and writers of the naturalistic image. The ability to counterfeit the appearance of a living thing meant that not only could the image explain a text, but it could also stand for the object itself. The use of the naturalistic image as a medium of communication was spurred not only by the need of the doctors and apothecaries to exchange information about sources of potent new cures, but also by the demand of the great compilers such as Gesner and Aldrovandi for the immutable image that could be recognized and used for identification of animals and plants that had dozens of different appellations. The problems inherent in the transposition of naturalistic image to printed reproduction may have limited somewhat the usefulness of these 'living images,' but the convention of the inclusion of the image with text in natural-history books was firmly established by the end of the sixteenth century.

By the beginning of the seventeenth century, nature was beginning to be divided and categorized. Sixteenth-century authors had prepared works on fishes and mammals, plants and insects, reptiles and minerals. The natural world began to be seen as distinct from civilization. Though the habits of mind that had made animal tales a part of human experience persisted in the emblem-books and to a certain extent in the works of the pandect authors, the idea of the classification of plants and creatures as particular entities had taken hold. What also began to emerge in the seventeenth century was a new perception of the land. Du Creux writes in 1664 that many had questioned how he could write about a country so featureless and devoid of landmarks as North America must be. Landmarks were cities, palaces, aqueducts – historic structures created by men and women that brought order to the landscape. The illustrations in the *Nuremberg Chronicle* do not show forests or rivers, but cities and bridges, tall towers and well-constructed harbours. But just as the map-makers had used animals to mark the land, the authors of the early eye-witness accounts began to demand drawings of natural wonders, of stupendous waterfalls and animal constructions, such as beaver dams, as well as of the great woods themselves. That the illustrators and gravers had difficulty in realizing a naturalistic veracity in the depiction of landscape should not be surprising, considering that the idea of the countryside as a subject for art was still far from prevalent in the Old World. The portrayers of the New World adopted the modes in which the Old World had expressed ideas of landscape. They created parks and hunting grounds, and grappled with the concepts of vast distances and strange vegetation by incorporating the by now familiar animals and plants drawn by the pandect illustrators of an earlier generation. The deer park and the ordered garden blooming in sunflowers and corn, as well as the Arcadian landscape (for which there was some precedent), all were used to depict what Du Creux had called 'the horror and immensity of woods and prairies.'

The process by which the natural world and its inhabitants became divorced from entanglements with human history was gradual, but, towards the end of the seventeenth century, naturalists such as John Ray and Francis Willughby could express succinctly a new view of what was becoming a specialized area of study. Ray wrote in 1678 that, in his books, 'we have wholly omitted what we find in other authors concerning ... hieroglyphics, emblems, morals, fables, presages or aught else appertaining to divinity, ethics, grammar, or any sort of human learning; and present ... only what properly relates to natural history.'[3] Nature had no connection with 'human learning'; it was other, and the natural order revealed more clearly than that of the human, the hand of God. By careful observation and meticulous measurement, and by portraying strictly what could be seen, the universe as it was created would be uncovered to human view. This separation between the constructions of men and the works of God meant that the landscape and the animals and plants in it must be classified anew. What were the affinities between living things that would reveal the mind of the Maker? This otherness and the inherent divinity of the natural world gave new significance to the work of the observers and illustrators of nature. Their strict attention to the visible elevated the importance of drawing as not only a counterfeit of nature, but a representation of essence. Michael Aaron Dennis, in a brilliant essay on

Robert Hooke's *Micrographia*, discusses how Hooke and other members of the Royal Society saw representation itself as a hermeneutic. For Hooke, 'seeing and representing *was* understanding.'[4] Thus the plates in *Micrographia* were 'central, rather than supplementary elements in the book.'[5]

At the same time, the new sense of the importance of observation and the pleasure taken by individuals in the picturing of natural beauty were reflected in the development of landscape illustration. The alteration in views about appreciation of the countryside that occurred during the eighteenth century led to a widespread taste for landscape painting, and, as Thomas notes, scenery became picturesque because it looked like a 'picture.' The documentation of a river, an unusual geological formation, a waterfall, or a mountainous view proceeded, however, not only from a desire to picture a scene, but also, as Bernard Smith suggests, from the same impulse that sought to depict clearly a specimen of plant or animal, as 'an assistance ... to truth.'[6] The idealized and organized landscapes supplied by de Bry and the other seventeenth-century engravers were replaced by views 'taken on the spot' that frequently included examples of the local flora and fauna. The notion of habitat, of a landscape inhabited by animals and plants, was not clearly articulated, but Thomas Pennant's interest in zoogeography was a precursor of the greater attention that would be devoted to this study in the nineteenth century. The notion of typical scenery, delineated by Smith in his study of the Cook expeditions, also informed the work of artists attempting to provide views of North America.

The Role of Drawing

That the standard or the 'truth' to which illustrations of the natural world were to conform changed over time is evident from the images included in this book. It would not, however, be correct to imply that artists learned how to copy nature better with each succeeding generation. Ligozzi and Weiditz prepared their drawings of plants with great fidelity, as did Georg Ehret two centuries later. Few have rendered a 'land Tort' better than John White, and who would wish to choose between the butterflies of Sibylla Merian and those of Moses Harris? How their precise and delicate renderings were communicated to the reader, however, has been the subject of much of this book. Problems and conventions of reproduction and the shifting emphasis created by context have meant that readers often saw the works of these artists very differently. In this final section, I would like to take up some of the threads concerning the role of images in both science and history and unravel them a little further. I have chosen to pay attention as much to images as to words, and this methodology has provided some unique insights and posed a number of problems. The use of visual materials as the ground for analysis is important and, as Martin Rudwick suggested, both necessary and timely. It was Rudwick's contention nearly twenty years ago that the study of illustrations in the history of science had been ignored by most historians for very unscientific reasons. He suggested that, 'in the hierarchy of our educational institutions, visual thinking is simply not valued as highly as verbal or mathematical dexterity ...' As a result, there exists 'the common but intellectually arrogant assumption that visual modes of communication are either a sop to the less intelligent or a way of pandering to a generation soaked in television.'[7]

Rudwick's contemporaries were not, however, the first to make an 'intellectually arrogant assumption' about the use of images. Mary G. Winkler and Albert Van Helden suggest that the Renaissance astronomer Galileo also made a similar assumption about the value of illustration. Despite the fact that both anatomy and botany had become visual disciplines in the sixteenth century, astronomy was still very much linked to mathematics, relying on the word, not on the image. Although Galileo used naturalistic representations in his early works, *Sidereus nuncius* (1610) and his *Letters on Sunspots* (1613), he included only diagrams in later works. Winkler and Van Helden attribute Galileo's reluctance to include images to his desire for prominence at court, and the urge to distance himself from the appellation of artisan. Telescope-makers and artists worked with their hands, and their status was lower than that of the scholar. Even in the earlier texts, Galileo had 'sacrificed the accuracy of the visual representations (which were, finally, visual aids) to the demands of text, which carried the real – and accurate – message. To Galileo one picture was not necessarily worth a thousand words.'[8] This ambivalence concerning the use of images reflects the criticisms levelled by Pliny in the first century at artists and copyists who could never realize a flower as it was throughout its life, nor accurately reproduce the best image (see above, chapter 3). We have noted it as well in Hieronymus Bock, but we have also observed the defence of the value of the pictorial image by Renaissance scholars such as Gesner[9] and Agricola,[10] and later by Catesby and Edwards, and the ready acceptance of the image as counterfeit by eighteenth-century naturalists like Linnaeus and Banks.

It is important to note, however, that the techniques of naturalistic representation first developed in the Renaissance, and to which later artists were heir, could be used to dissimulate as well as to record. As Martin Kemp points out, naturalistic representation is 'a distinctly double-edged sword. An imaginary animal can be depicted with a credibility no less sharp than an existing one.'[11] Francis Haskell's discussion of the haphazard use of portraits by engravers and publishers is reflected in the words of Daniel Defoe, who on a visit to Holyrood House in 1726, examined the portrait gallery: 'The North Side is taken up with one large Gallery, reaching the whole Length of the House, famous for having the Pictures of all the Kings of Scotland ... But, in my opinion, as the Pictures cannot be, and are not supposed to be Originals, but just a Face and Dress left to the Discretion of the Limner, and so are all Guess-work, I see no Rarity, or, indeed, any Thing Valuable in it.'[12] Gombrich has taken this understanding even further. 'Pictures,' he writes, 'cannot assert. While a verbal account need leave us in no doubt that it claims to describe an existing state of affairs, the uncaptioned pictorial representation may just as easily refer to an existing building as to a memory, a plan or a fantasy.'[13]

The dual nature of naturalistic representation may perhaps explain the widespread and recurrent mistrust of illustration, which has led to a suspicion, at least in some circles, of the information value of the picture. Neil Harris, in a thoughtful analysis of the 'halftone revolution' of the late nineteenth century, notes that contemporary critics complained that the easy availability of photographic reproductions had led to a 'rage for illustration,' which would in time undermine the truth value of the text. Harris quotes C.F. Tucker Brooke, who wrote to the editor of *The Dial* to alert readers that 'pictures irresponsibly selected, and inserted without adequate investigation, can easily lead to more serious misapprehension than would result from glaring error in the letter-press.'[14] Not only might a picture lie, but it might lie more forcibly than text. A 1911 editorial in *Harper's Weekly*, which had used half-tone illustrations since the 1880s, suggested that illustrations improperly used became 'a mental drug.' The editorial continued: 'It would be safe to say that a young mind, overfed pictorially, will scarcely be likely to do any original thinking.'[15] Harris has attributed to the lingering echo of these old debates some of the reluctance of his colleagues – contemporary intellectual historians – to engage themselves in the analysis of the impact of the 'halftone revolution.' Much as Rudwick had suggested in respect to historians of science, intellectual historians, '[a]s students of the word, with a large investment in careful verbal analysis ... have tended to deprecate surrogates thrown up the Industrial Revolution, surrogates that threatened the primacy of printed communication and menaced the very concept of authenticity itself.'[16]

Art historians, on the other hand, have never been reluctant to engage the image. From their researches and new multidisciplinary studies have come some of the freshest insights into the significance of the image and its role in the history of thought. In 1938, William Ivins asserted that 'sight has today become the principal avenue of the sensuous awarenesses upon which systematic thought about nature is based.'[17] His insistence on the importance of the role played by visual images in the creation of science has been incorporated in the works of some contemporary authors concerned with the genesis and mechanisms of scientific thought. Eugene Ferguson contends that, to understand the development of Western technology, we must appreciate non-verbal thinking. The heavily illustrated codices of the fifteenth and sixteenth centuries relied on illustrations, not text, to transmit knowledge about mechanical and technological processes and machinery. The encyclopedists of the eighteenth century also recognized the value of illustrations when recording the work of the ateliers of France. Despite a tradition of pictorial representation in engineering history, Ferguson, like Rudwick, acknowledges that, 'because perceptive processes are not assumed to entail "hard thinking," it has been customary to consider nonverbal thought among the more primitive stages in the development of cognitive processes and inferior to verbal and mathematical thought.'[18] Historians of science, such as Robert Scott Root-Bernstein and Michael Lynch, are also beginning to examine the role of representation in scientific thinking. Root-Bernstein's work with scientists has suggested to him that 'one thing seems certain. Most eminent scientists agree that nonverbal forms of thought are much more important to their work than verbal ones.'[19] Lynch observes that, in the development of scientific ideas, 'in many cases there is no way to compare a representation of a biological phenomenon to the "real" thing, since the thing becomes coherently visible only as a function of the representational work.'[20] Thus, not only do scientists think in pictures, but in some cases their thinking must take the form of a picture before it can be understood and communicated at all. Representation, as Robert Hooke understood, becomes hermeneutic.

It is the art historian Barbara Stafford, however, with her interests in the conjunction of representation and knowledge, who has perhaps best understood the historical significance of this new kind of vision. For her the advent and widespread use of electronic media with its graphical interfaces and easy access to imagery marks a renaissance of 'visual aptitude,' eclipsed in the modern era. She, too, acknowledges the scant attention paid to the 'mind-shaping powers of ocular, kinesthetic, and auditory skills ... scarcely articulated in the tale of Western civilization's turn to the cultivation of the interior ...' The rediscovery of visualization is for her at the core of a new understanding of the history of science and art:

> Uncovering this lost epistemological dimension of the informed and performative gaze, and with it the complex interface of early modern nature and artifice revealed in moments of enlightening recreation, seems all the more important in our computer era. Now old intellectual traditions based on crayons, loose-leaf paper, and paste are also being replaced by playful high-tech tools and visually appealing programs. The rise of electronic media casts print culture, as well as the histories of art and science on which these disciplines are grounded, in sharp relief.[21]

She goes on to advocate a new approach to the study of pictorial representation:

> We need, therefore, to get beyond the artificial dichotomy presently entrenched in our society between higher cognitive function and the supposedly merely physical manufacture of 'pretty pictures.' In the integrated (not just interdisciplinary) research of the future, the traditional fields studying the development and techniques of representation will have to merge with the ongoing inquiry into visualization. In light of the present electronic upheaval, the historical understanding of images must form part of a continuum looking at the production, function, and meaning of every kind of design.[22]

One of the very few to look at the process involved in the transformation of information into visual forms has been Edward Tufte. 'Envisioning information' is the term used by Tufte to describe the process by which complex data are rendered into a visual analogue. He observes that 'all communication between the readers of an image and the makers of an image must now take place on a two-dimensional surface. Escaping this flatland is the essential task of envisioning information – for all the interesting worlds (physical, biological, imaginary, human) that we seek to understand are multivariate in nature.'[23] Tufte has examined primarily the graphic representation of three-dimensional space (maps) and time (schedules, astronomical observations) in two-dimensional media, but the information concerning classification embodied in an Ehret drawing is certainly as complex as a railway timetable. The conventions that were adopted by Gesner, de Bry, the illustrators of florilegia, and the artist amateurs who worked so diligently for the great European collectors, enabled them to bring before the chief human sense the critical data required for understanding what were, in many cases, new phenomena. Current research into visualization confirms that, as the amount of information increases, the need for quick apprehension of large amounts of data makes visualization a key tool for achieving insight into the behaviour of systems and the nature of the world.[24] If this observation is accepted, one can look at Ivins's statement that images were necessary to the development of science with new interest.[25] This current inquiry is but a partial and preliminary attempt to come to grips with an integrated study of images in light of new thinking about the importance of the visual in the understanding of phenomena. What my examination of this set of images has revealed, however, is that, if images are to be studied with care and attention, there are three different aspects of the study of representation that demand comment.

Chronology

Martin Rudwick, in another work in the history of science, *The Great Devonian Controversy: The Shaping of Scientific Knowledge among Gentlemen Specialists* (1985), points out the importance of strict chronology:

> If scientific knowledge is to be studied *in the making*, the closest attention

> must be paid to strict chronology, not only in description but also in analysis ... The risk is that the description and analysis may be irreparably distorted by the historian's or the reader's knowledge of the outcome of the episode or the 'correct' solution of the controversy. Narrative in the service of understanding the shaping of knowledge must rigorously and self-consciously avoid hindsight.[26]

Just as visual thinking has been seen as a less precise version of verbal thought, so have images been used by historians in a far less rigorous manner than texts. Images have been treated as 'illustrative material,' added to text more often for decoration than for illumination. This tendency has been compounded by the practice of museums that tend to arrange materials in only vague chronologies, positioning them for aesthetic resonance rather than intellectual coherence.[27] Rudwick's understanding of strict chronology means that images, as well as texts, must be seen in relation to their position in time. History, after all, is concerned with what went before and what came after. Rudwick, like Foucault, however, warns against the tendency to see outcomes, to posit directional relationships on the material. This is a strong temptation when examining naturalistic representation, since, with hindsight, we ask, why is this image distorted? When do artists learn how to paint what is actually there, what we see? These were the questions with which this study began, and part of the work in this discourse has been to attempt to understand the image in its own context. David Knight more than once observes that 'there is no proper way of drawing any animal, and two excellent pictures of the same creature may look very different ... We do not therefore simply find progress in zoological illustrations ... Indeed to approach the history of zoology through illustration, is a ready way of dropping the idea of science as cumulative progress to indubitable truth based upon some "scientific method."'[28] An appreciation of chronology, of sequence, has informed this work. Placing all images in the correct order has made certain patterns visible. In some cases, these patterns allow us to make hypotheses about broad discontinuities or displacements of one system of thought by another. Thus, we can see that the use of emblematic animals as signs or marks of countries disappears, but, at the same time, the animal as mark becomes transmuted to what Ford has called the 'reference image,' the way the animal looks as the definition of itself. What is significant about these reference images is their longevity. As Foucault has noted, 'the same, the repetition, and the uninterrupted are no less problematic than the ruptures.'[29]

What can we understand about the repeated use over time of a particular image? Dürer's drawing and woodblock print of the rhinoceros is so convincing that it becomes accepted as an image of the real thing. It is satisfying as a work of art, and its value as accurate portrayal was likely enhanced by Dürer's reputation as a painter of animals and plants. In addition, the original cut is accompanied by text – sometimes reproduced in subsequent re-engravings, but often omitted – that adds the veracity of words to the image. Not only does the animal look like this, it looks like its description says it should look. For both Dürer and his audience, then, there was no dissonance between the pictorial image of the rhinoceros and the rhinoceros itself. This reference image fixed for over two-and-a-half centuries how the rhinoceros should look. What does this repetition of image, this fixity of representation, imply for the understanding of the making of scientific knowledge? Elizabeth Eisenstein has maintained that typographical fixity was 'a basic requirement for the rapid advancement of learning,'[30] since errors could be corrected in subsequent editions. While this may well apply to text – and she maintains it applies as well to images, such as the illustrations for Vesalius's second edition – we should not ignore the lesson of Dürer's rhinoceros. The typographical fixity of the image has become an iconographical fixity; in the age of Linnaeus, the Renaissance rhinoceros with the 'hornlet' remains an emblem of a species.

Dürer's representation of the rhinoceros agreed with its textual description, but we have seen how the Hennepin reference image of Niagara, first issued in 1697, became the iconographic template for 'eye-witness' description until well into the nineteenth century. This rupture between text and image requires further exploration. J.A. Lohne has examined 'The Increasing Corruption of Newton's Diagrams.' Despite the fact that Newton himself executed the original drawings, which documented his experiments with care in the manner expected of members of the Royal Society, the artists and editors who prepared the printed texts paid less than adequate attention to their reproduction. As a result, Lohne points out that, once engraved, the diagrams were seen as reference images, reused in subsequent editions without correction, even though 'their diagrams [are] often violating fundamental optical laws.'[31] John Levene examines several editions of Descartes's *La dioptrique*, noting that the image reproduced in 'modern opthalmic literature' is incorrect, being reproduced from an 1824 French edition with a diagram that distorts the original and correct engraving prepared under Descartes's supervision.[32] Root-Bernstein has observed that 'visual and other nonverbal forms of thinking profer to philosophers vast wildernesses in need of exploration,'[33] and it is obvious as well that conclusions drawn about the importance of typographical fixity by an examination of text alone must be revised.

Dominant Purpose

If hindsight is to be avoided in examining the role of the image, even more important is to examine what Gombrich calls the dominant purpose of the image:

> the great variety of styles we encounter in the images of past and present civilizations cannot be assessed and interpreted without a clear understanding of the dominant purpose they are intended to serve. It is the neglect of this dimension which has suggested to some critics that the range of representational styles must somehow reflect a variety of ways in which the world is seen. There is only one step from this assumption to the assertion of a complete cultural relativism which denies that there are standards of accuracy in visual representation because it is all a matter of convention.[34]

The issue of dominant purpose becomes extremely significant when examining the use of naturalistic representations. How else to explain in the age of Ligozzi the appearance of the su and the simivulpa? Martin Kemp has also explored the intersection between natural history and art, and points out the utter dependence of the viewer of an image on 'prior knowledge, automatic expectation, illustrative technique, emotional context and the given framework of verbal information, if we are able to read an image in a meaningful way.'[35] Comparing the exquisite watercolour portrait of a mandrake plant by Ligozzi, with an engraving of the same plant 'humanized' by Abraham Bosse, Kemp observes that 'Bosse was not simply being stupid. His representation was locked into a system of beliefs in which form and meaning might be regarded as inseparable – a system in which accidental configurations of mandragora roots could indeed act as an effective talisman, with or without the assistance of a human sculptor.'[36] It was not that Bosse did not know how mandrakes looked in the wild, or that he could not, if he turned his hand to it, execute a competent engraving of the plant (the flowers and leaves are lifelike), it was that the representation of the plant as a portrait was not his purpose, as it had been Ligozzi's. Accuracy in naturalistic representation might be subordinated to concern for style, for viewer's expectations, or, as in de Bry's case in some of the engravings in *America*, with a concern to show 'all the facts' in a single composite image.

It is at this interface between purpose and representation that one can begin to explore the notion of what Foucault has referred to as 'savoir' – not what is known but the domain in which knowing takes place. Style is in effect a reflection of *savoir*. How the artist depicts information, how the data are configured visually, is intimately related not simply to the content, but to the ability to control and direct the content. To 'know how' is to obtain mastery, and in the field of visual representation (and one suspects in text), *savoir* is linked to the ascription of meaning. I have configured the information in this way because it has meaning for me. Thus, the New World resembles a deer park because not only is it a familiar and recognizable image, but it is also an expression of order and mastery. Similarly, Davies and Edwards can turn waterfalls and birds into specimen portraits because classification is a means of knowing the order of the universe. Edwards in his petition to God does not want to gain more information (connaissance); rather, he wants to lose all desire for natural history, so that he 'may become an intelligent spirit' which may 'see and know how the parts of the great Universe are connected with each other ...'[37] In the material-history analysis of the image, then, style is not simply the external shell under which true meaning is hidden; it is in itself redolent with the artist's understanding of meaning.[38] Understanding and analysing style in images is as significant for an understanding of 'the making of science' as close reading of text.

Context

Context means 'with text,' and the final aspect that requires some comment in conclusion is that of the contextual appearance of the images examined. Modern methods of reproduction – the 'halftone revolution' – have made it difficult for contemporary viewers and readers to appreciate the importance of the context in which the image is physically presented. Elizabeth Eisenstein commented on the rupture between text and image which occurred when hand-illuminated manuscripts were replaced by printed volumes. With rare exception, the text and image no longer interpenetrated each other, supporting and informing each other. The 'copper-plate revolution' completed this rupture, making it simpler to print letterpress separately from image, and one has only to think of perhaps the greatest illustrated book of the nineteenth century, Audubon's *The Birds of America*, which was published without text (the text was printed later as a separate volume), to realize how much the rift had widened. Too often, scholars examine images bereft of their context. I have tried, whenever possible, to examine an original edition, or at least a fascimile reproduction, so that I was able to understand the often complex relationship between text and image. The repetition of images by de Bry is as significant for what it has to tell us about the reader's demand for veracity as for what it says about the publisher's economies. Most artists and engravers did not expect to see their images stand alone, and they worked to the demands of author and publisher. Ehret's watercolours are what they are because they are used to illustrate a particular kind of text. Even where no printed text exists, as in the Newfoundland watercolours, Ehret insured that his images corresponded not only to the dried plant fixed on the sheet, but to the herbarium notes. These works, then, cannot be seen, though they often are, as 'works of art' that answer to an inner aesthetic of the artist. As pleasing to the eye as they may be, these images answer to the needs of the context in which they were created, a context of words, specimens, and specialized understanding. Even Pierre Desceliers, whose manuscript map was destined to decorate the wall of a princely chamber, attended to the details of context, placing, as did the other cartographers studied by Wilma George, the correct animals on his map.

Those who would study images as evidence must also be conscious of the artefacts of reproduction, or what Peter Taylor and Ann Blum call printing features: 'We would argue that "printing features" are not separable from the meanings of graphic representation.'[39] Looking at images reproduced in modern books as black-and-white halftones destroys perceptions of scale, texture, and colour. There is something truly amazing in turning the leaves of a book the size of Aleksandr Filippovich Postels's (1801–1871)*Illustrationes Algarum in itinere circa orbem jussu imperatoris Nicolai I* (1840). Its very massiveness (69.0 centimetres in height) forces one to confront the issue of the use of these books, the readership, the whole economy which supported publications of this nature. By contrast, handling the sketchbooks of George Back, which survived the vicissitudes of his arctic journeys, provides information about the way in which the artist saw the landscape and its inhabitants and the way in which he composed it to give it personal meaning. Similarly, the 'feel' of old paper, the observation of how the ink sets on it, or the way in which colour either sinks into or stands up on the page are important considerations in making sense of the complex interrelationship between the technology of printing and the nature of representation. What is possible as a watercolour has not always been possible in print, as George Edwards found to his dismay. Finally, we have noted that colour, for so many centuries achieved with such great pains, is not a trivial matter. Problems of accurate colour reproduction plagued artists such as de Passe, Merian, Catesby, and Edwards, and the whole history of the colouring trade, which involved large numbers of women and young children, has been relatively little explored. Understanding the presence or lack of colour, then, means being aware of the social and economic conditions of the time and the purposes of the artist or author.

Cognitive Anchorage

Given some of the difficulties in studying images noted above, why should the analysis of representation form part of historical enquiry? Doesn't the analysis of text supersede the analysis of image, since text is often the descriptor and interpreter of the image? While acknowledging that the examination of visual representation is a neglected part of historical discourse, is there a need to privilege images for special consideration? I would like to suggest, as some of those cited above have observed, that visual thinking is of a different order from verbal thinking, and that the understanding of represenation is an entrée to an understanding of a different order. What is the function of the image in thought and, more particularly, in scientific thought? While this too brief survey cannot presume to examine this question in a more than cursory way, there are certain indications that images have a unique function in the creation of knowledge.

It has long been acknowledged, particularly by those examining the work of scientists, that diagrams, doodlings, visual symbols, and the like have an important function in the creative process. In a discussion on the eliminability of scientific diagrams, James Griesemer poses this question: 'But are diagrams mere scaffolding to express ideas as conveniently as possible or are they fundamental to the practice of science?'[40] He answers his own question by suggesting that, while diagrams which have been used to formulate concepts might be thrown away once the concept itself is reformulated in sentences, 'even if diagrams are logically eliminable in a formal reductionist sense they can nevertheless be heuristic in a strong sense as well.'[41] The idea of representation as heuristic returns us to Hooke and *Micrographia* and to Edwards. The thing must be seen before it can be known.

This sense that representation permits interpretation is part of Michael Lynch's careful analysis of the relationship among diagrams, photographs, and digital imagery in science,[42] and the foundation of the notion that visualization of complex data provides not only understanding, but also mastery. While more work needs to be done, it is tempting to postulate that the use of images as information occurs at points when the data are too complex for simple verbal transcription. How was de Bry to explain in words the actual appearance of the New World? What fixed images could European readers use to understand the appearance of corn, sunflowers, turkeys, or Native peoples? Only by representing the manuscript drawings of eye-witness observers could de Bry hope to convey the true strangeness, the otherness of America. For the early scientific botanists, who had no verbal agreement on the definition of sepal, pericarp, or stamen, the representation was the necessary condition for knowledge about the plant and its uses. By encoding complex data in visualizations, scientists are able to record new knowledge in such a way that incremental discovery can be grafted easily to the base of the known. Thus, Hunter and his colleagues take Stubbs's painting with them in their first essay into comparative anatomy of the moose. The heroic enterprise of eighteenth-century science, which Bernard Smith describes, consisted primarily in making a visual record of the known world. Once the record was mostly complete, new ways of looking at the world, such as the theoretical approaches of Humboldt, could be developed on a firm foundation of visualized understanding. Sir Ernst Gombrich refers to our consistent perception of the visual world as 'the cognitive anchorage we need in our effort after meaning.'[43] I would like to suggest that the image, the visual representation of the thing, is indeed at the foundation of much of our cognition of the world. Vision is our primary sense, and it is through visual apprehension that we attempt to comprehend the order of things. To make sense of the ways in which knowledge has been formed and reformed, the examination of its visualization over time is worthy of study.

NOTES

Introduction: *The Bittern from Hudson's-Bay*

1 James Isham, *James Isham's Observations on Hudson's Bay and Notes and Observations on a book entitled* A Voyage to Hudson Bay in the Dobbs Galley, 1749 (London: Hudson's Bay Record Society, 1949), 321.
2 *Some Memoirs of the Life and Works of George Edwards, Fellow of the Royal and Antiquarian Societies* (London: Printed for J. Robson Bookseller, 1776), 6–7.
3 Linnaeus to C.J. Trew, quoted in Gerta Calmann, *Ehret, Flower Painter Extraordinary: An Illustrated Life* (Oxford: Phaidon, 1977), 97.
4 Quoted in A. Stuart Mason, *George Edwards: The Bedell and His Birds* (London: Royal College of Physicians, 1992), 48.
5 In a study of this nature, examining originals rather than reprinted images is critical. Where possible, given the constraints of geography and time, I have tried to examine the original works. I comment more fully in the conclusion on the significance of the physical attributes of the manuscripts, prints, and books as aids to understanding the role of the image.
6 Martin J.S. Rudwick, 'The Emergence of a Visual Language for Geological Science, 1760–1840,' *History of Science* 14 (1976): 149–95.
7 See, however, I. Bernard Cohen's *Album of Science: From Leonardo to Lavoisier, 1450–1800* (New York: Charles Scribner's Sons, 1980).
8 The Cumming and Quinn series on the discovery and exploration of North America, while extremely valuable for the number of images it portrays, is a particularly glaring example of this practice. This use of images is equivalent to the exhibition of 'breakers' in the book and print trade, illustrations torn from texts and placed on sale or display for their aesthetic or curiosity value.
9 Brian Baigrie, ed., *Picturing Knowledge: Historical and Philosophical Problems Concerning the Use of Art in Science* (Toronto: University of Toronto Press, 1996), xviii.
10 Robert S. Root-Bernstein, 'Visual Thinking: The Art of Imagining Reality,' *The*

Visual Arts and Sciences: Transactions of the American Philosophical Society 75, pt. 6 (1985): 50–67.

11 Helen B. Mules, *Flowers in Books and Drawings, ca 940–1840* (New York: J. Pierpont Morgan Library, 1980); Edward Nygren, *Views and Visions: American Landscape before 1830* (Washington, DC: Corcoran Gallery of Art, 1986); Hugh Honour, *The European Vision of America: A Special Exhibition to Honor the Bicentennial of the United States ...* Cat. no. 92 (Cleveland: Cleveland Museum of Art, 1975); Larry Barsness, *The Bison in Art: A Graphic Chronicle of the American Bison* (Fort Worth, TX: Northland Press and the Amon Carter Museum, 1977); Susan Danforth and William H. McNeill, *Encountering the New World, 1493 to 1800* (Providence RI: John Carter Brown Library, 1991); Jay Levenson, ed., *Circa 1492: Art in the Age of Exploration*, Cat. no. 205 (Washington, DC: National Gallery of Art; New Haven, CT, and London: Yale University Press, 1991); Brian J. Ford, *Images of Science: A History of Scientific Illustration* (London: The British Library, 1992); National Library of Canada, *Passages: A Treasure Trove of North American Exploration/Passages: Un écrin des explorations de l'Amérique du Nord* (Ottawa: National Archives of Canada, 1992).

12 'Early Italian Nature Studies and the Early Calendar Landscape,' *Journal of the Warburg and Courtauld Institutes* 1/13 (1950): 13–47.

13 Francis Klingender, *Animals in Art and Thought to the End of the Middle Ages* (London: Routledge & Kegan Paul, 1971); Svetlana Alpers, *The Art of Describing: Dutch Art in the 17th Century* (Chicago: University of Chicago Press, 1983); Barbara Maria Stafford, *Voyage into Substance: Art and Science and the Illustrated Travel Account, 1760–1840* (Cambridge, MA: MIT Press, 1984); Hugh Honour, *The New Golden Land: European Images of America from the Discoveries to the Present Time* (New York: Pantheon, 1975); Bernard Smith, *European Vision and the South Pacific*, 2d ed. (New Haven, CT: Yale University Press, 1988).

14 S. Peter Dance, *The Art of Natural History: Animal Illustrators and Their Work* (Woodstock, NY: Overlook, 1976); A.M. Lysaght, *The Book of Birds* (London: Phaidon, 1975); Peyton Skipwith, *The Great Bird Illustrators and Their Art, 1730–1930* (New York: A&W Publishers, 1979); S. Sitwell, H. Buchanan, and J. Fisher, *Fine Bird Books, 1700–1900* (New York: Atlantic Monthly Press, 1990).

15 Jean Anker, *Bird Books and Bird Art* (Copenhagen, 1938; repr. New York: Arno, 1974); C.E. Raven, *English Naturalists from Neckham to Ray: A Study of the Making of the Modern World* (Cambridge: Cambridge University Press, 1947); Wilfrid Blunt, *The Art of Botanical Illustration*, 4th ed. (London: Collins, 1971); Agnes Arber, *Herbals, Their Origin and Evolution: A Chapter in the History of Botany, 1470–1670* (1938; 2d ed., Darien, CT: Hafner, 1970); D.M. Knight, *Natural Science Books in English, 1600–1900* (London: B.T. Batsford, 1972), and *Zoological Illustration: An Essay Towards a History of Printed Zoological Pictures* (Folkestone: Dawson Archon Books, 1977).

16 Francis Haskell, *History and Its Images: Art and the Interpretation of the Past* (New Haven, CT: Yale University Press, 1993), 2.
17 Ibid., 9–10.
18 William M. Ivins, Jr, *Prints and Visual Communication* (Cambridge, MA: MIT Press, 1985), vii.
19 Ibid., 3.
20 Edward Tufte, *Envisioning Information* (Cheshire, CT: Graphics Press, 1990).
21 Ulrich Land and Michel Grave, 'Data Structures in Scientific Visualization,' in *Focus on Scientific Visualization*, H. Hagen, H. Müller, and G.M. Nielson, ed., (Berlin: Springer-Verlag, 1993), 85.
22 Stephen Lubar and W.D. Kingery, *History from Things: Essays on Material Culture* (Washington, DC: Smithsonian Institution Press, 1993), ix.
23 The manner in which the material historian works is perhaps best understood in the field of archaeology. In a famous example, an American historical archaeologist began to see the pattern in the distribution of broken pipe-stems in front of the foundations of an early New England building. There were no records of the building's use, no textual evidence of its purpose, but by reading the material evidence James Deetz was able to hypothesize that a building where broken pipe-stems lay scattered might well be a store, a place where people gathered to smoke their long clay pipes, breaking off the pipe-stems as they became worn: James Deetz, *In Small Things Forgotten: The Archaeology of Early American Life* (New York: Doubleday Anchor, 1977).
24 Jules David Prown, 'The Truth of Material Culture: History or Fiction?' in *History from Things*, ed. Lubar and Kingery, 4.
25 Works of art can, of course, be analysed as artefacts. Prown (ibid.) also notes that, for some art historians, such as Irving Lavin of Princeton, all things made by people are art – 'art is equatable with artifacts.'
26 Michel Foucault, *The Archaeology of Knowledge*, trans. A.M. Sheridan Smith (London: Tavistock, 1972), 28.
27 Ibid., 193–4.
28 Henry Glassie, *Folk Housing in Middle Virginia* (Knoxville: University of Tennessee Press, 1975), vii.
29 Stephen Greenblatt, *Marvelous Possessions: The Wonder of the New World* (Chicago: University of Chicago Press, 1991), 7.
30 Knight, *Zoological Illustration*, 39.
31 Foucault, *Archaeology of Knowledge*, 174.

Chapter 1: Emblematic Animals

1 William Brandon, *New Worlds for Old: Reports from the New World and Their Effect on the Development of Social Thought in Europe, 1500–1800* (Athens: Ohio University Press, 1986), 7.
2 Suggested by Susan Danforth, in Susan Danforth, and William H. McNeill,

eds., *Encountering the New World, 1493 to 1800* (Providence, RI: John Carter Brown Library, 1991), fig. 2. The use and reuse of blocks to represent a place by type, that is, the idea of 'city,' is obvious in the *Nuremberg Chronicle*, also published in 1493. Here, the cities of Neapolis (Naples), Lyon, and Mainz, as well as others, are all represented by the same image. For discussion of the repetition of images, see chapter 2.

3 Hugh Honour, *The New Golden Land: European Images of America from the Discoveries to the Present Time* (New York: Pantheon, 1975), 6–7.

4 Ibid., 5.

5 Quoted in ibid.

6 Stephen Greenblatt, *Marvelous Possessions: The Wonder of the New World* (Chicago: University of Chicago Press, 1991), 76–8. The long-tailed, colourful New World parrots seem particularly to have been prized, and were brought back to Europe in March 1494, where they quickly made their appearance on the American section of a world map (the Cantino map) created for the Italian ambassador of the Duke of Ferrara in Lisbon in 1502. Green African parrots, often referred to as 'popinjays,' were already familiar domesticated birds in Europe and were featured in medieval manuscripts and illustrated prayer books of the fourteenth century (G. Evelyn Hutchinson, 'Attitudes toward Nature in Medieval England: The Alphonso and Bird Psalters,' *Isis* 65 [1974]: 11.) They are shown on the African section of the same 1502 map.

7 Quoted in Honour, *The New Golden Land*, 12.

8 Quoted in W.P. Cumming, R.A. Skelton, and D.B. Quinn, *The Discovery of North America, 1630–1776* (New York: American Heritage, 1971–2), 53.

9 Recorded in the Household Book of Henry VII, and noted in Samuel Eliot Morison, *The European Discovery of America: The Northern Voyages, AD 500–1600* (New York: Oxford University Press, 1971), 219.

10 Quoted in Cumming et al., *The Discovery of North America*, 53.

11 Morison, *The European Discovery*, 210.

12 Quoted in Douglas R. McManus, *European Impressions of the New England Coast, 1497–1620*, Research paper no. 139 (Chicago: University of Chicago, Department of Geography, 1972), 63.

13 Jacques Cartier, *The Voyages of Jacques Cartier*, ed. Ramsay Cook (Toronto: University of Toronto Press, 1993), 17–18.

14 Ibid., 100.

15 Quoted in ibid., 57–8.

16 Quoted in Greenblatt, *Marvelous Possessions*, 78.

17 R.A. Skelton, *Explorers' Maps: Chapters in the Cartographic Record of Geographical Discovery* (London: Routledge & Kegan Paul, 1960), 325.

18 W.F. Ganong, *Crucial Maps in the Early Cartography and Place-Nomenclature of the Atlantic Coast of Canada* (Toronto: University of Toronto Press, 1964), 176.

19 François-Marc Gagnon and Denise Petel identify these whalers with the Beothuk Indians of Newfoundland. See their discussion in *Hommes effarables et*

bestes sauvaiges: Images du Nouveau-Monde d'après les voyages de Jacques Cartier (Montreal: Les Éditions du Boréal Express, 1986), 158–9.

20 Quoted in Wilma George, *Animals and Maps* (Berkeley: University of California Press, 1969), 91. Desceliers depicted a turkey on his 1550 map.

21 'The only earlier bison is, very dubiously, the tapaca of Desceliers' 1546 map': ibid., 94.

22 Ibid., 160.

23 Ibid., 91.

24 Quoted in Cartier, *Voyages*, 5.

25 Ganong, *Crucial Maps*, 195.

26 Razor-bill auks: Cartier, *Voyages*, 5.

27 Ibid., 131. While this strange fauna is never heard from again, there appear in an early-seventeenth-century allegory of America by Crispijn de Passe two winged demons with long tails who hover above an altar presided over by an enthroned winged female, which Hugh Honour suggests is a depiction of Aztec sacrificial rites: Honour, *The New Golden Land*, 88, fig. 80.

28 For a discussion of the rabbit in maps, see George, *Animals and Maps*, 93–4.

29 Cartier, *Voyages*, 87.

30 Cartier remarks that the Native people he met in the Gaspé were the 'sorriest folk there can be in the world ... They have no other dwelling but their canoes, which they turn upside down and sleep on the ground underneath': ibid., 24–5.

31 Ibid., 15 and 46.

32 Ibid., 48.

33 Jessica Rawson, ed., *Animals in Art* (London: British Museum Publications, 1977), 128.

34 Johannis de Laët, *Novus orbis, seu Descriptiones Indiae Occidentales, libri XVIII* (Leyden, 1633), 38.

35 Swift, quoted in George, *Animals and Maps*, 21.

36 Ibid., 23.

37 Ibid., 25.

38 Cartier, *Voyages*, 127.

39 Thevet also notes that 'several plants and bushes were brought back from there which today are to be seen in the royal garden at Fontainebleau': *La Cosmographie universelle*, in *André Thevet's North America: A Sixteenth-Century View*, ed. and trans. Robert Schlesinger and A.P. Stabler (Kingston: McGill-Queen's University Press, 1986), 48.

40 Albrecht Dürer had even drawn a lynx from life in 1521, during his visit to the beast-garden at the King's House in Brussels: illustrated in Jay A. Levenson, ed., *Circa 1492: Art in the Age of Exploration*, Cat. no. 205 (Washington, DC: National Gallery of Art; New Haven, CT, and London: Yale University Press, 1991), 299.

41 C.E. Raven, *English Naturalists from Neckham to Ray: A Study of the Making of the Modern World* (Cambridge: Cambridge University Press, 1947), 33.

42 Quoted in ibid., 2.
43 Quoted in Michel Foucault, *The Order of Things: An Archaeology of the Human Sciences* (New York: Vintage, 1973), 21.
44 Thevet, *Cosmographie Universelle*, in *André Thevet's North America,* 48.
45 Cartier, *Voyages*, 14, 74, 58.
46 Brian J. Ford, *Images of Science: A History of Scientific Illustration* (London: The British Library, 1992), note to 48.
47 Ganong, *Crucial Maps*, 122.
48 Ibid., 121–2.
49 Ibid., 122.
50 'Ie ne doute point, Lecteur, que la description de ceste presente histoire ne te mette aucunemēt en admiration, tant pour la varieté des choses qui te font à l'oeil demōstrées, que pour plusieurs autres qui de prime face te semblerōt plustost monstrueuses que naturelles. Mais apres avoir meuremēt cōsideré les grās effects de nostre mere Nature, ie croy fecement que telle opinion n'aura plus de lieu en ton esprit': André Thevet, *Les Singularitez de la France Antarctique, Autrement nommée Amerique: & de plusieurs Terres & Isles decouvertes de nostre temps* (Paris: Maurice de la Parte, 1558), np (my translation).
51 Thevet, *Cosmographie universelle*, in *André Thevet's North America,* 140.
52 Edward Topsell, *History of Four-footed Beasts and Serpents and Insects*, vol. 1 (1658; repr. New York: DaCapo, 1967), 511.
53 George, *Animals and Maps*, 75.
54 Quoted in Wilma George, 'Sources and Background to Discoveries of New Animals in the Sixteenth and Seventeenth Centuries,' *History of Science* 18 (1980): 89.
55 Thevet, 'Grand Insulaire,' in *André Thevet's North America,* 244. Albrecht Dürer notes several times in the diary he kept of his 1520–1 journey through The Netherlands that he acquired 'buffalo horns': noted in William M. Conway, ed., *The Writings of Albrecht Dürer* (London: Peter Owen, 1958), 102, 105, 115.
56 Thevet, 'Grand Insulaire,' in *André Thevet's North America,* 130. One wonders if the enmity between horse and bison is yet another retelling of the tale of the unicorn and the elephant. See below, p. 254.
57 Thevet, 'Grand Insulaire,' in *André Thevet's North America*, 88.
58 F.G. Roe, *The North American Buffalo: A Critical Study of the Species in Its Wild State*, 2d ed. (Toronto: University of Toronto Press, 1972), 860 and 206n.
59 The animal appears twice – once alone, and once with a parrot, identified as *Psittacus*, on its back.
60 Adrian Forsyth, *Mammals of the Canadian Wild* (Camden East, ON: Camden House, 1985), 328.
61 Quoted in Honour, *The New Golden Land*, 40.
62 Forsyth, *Mammals*, 330.
63 Topsell, *The History of Four-footed Beasts*, 16.
64 Hugh Honour, *The European Vision of America: A Special Exhibition to Honor*

the Bicentennial of the United States ... Cat. no. 92 (Cleveland: Cleveland Museum of Art, 1975).

65 William Ashworth, 'The Persistent Beast: Recurring Images in Early Zoological Illustration,' in *The Natural Sciences and the Arts: Aspects of Interaction from the Renaissance to the 20th Century. An International Symposium*, ed. Allan Ellenius (Stockholm: Almqvist & Wiksell International, 1985), 65.

Chapter 2: Naturalism and the Counterfeit of Nature

1 Fernandez de Oviedo y Valdés, *Historia general de las Indias* (1535), quoted in Hugh Honour, *The European Vision of America: A Special Exhibition to Honor the Bicentennial of the United States ...,* Cat. no. 92 (Cleveland: Cleveland Museum of Art, 1975), 1.

2 David Knight, *Zoological Illustration: An Essay towards a History of Printed Zoological Pictures* (Folkestone: Dawson Archon Books, 1977), 17.

3 Pliny noted that elephants were first seen in Italy in 280 B.C., and Matthew Paris, a monk of St Albans in England, did manage to draw a very recognizable elephant from life around 1255. This elephant was a gift to Henry II from Louis IX of France and was housed in the Tower. In England, the Tower became the home for a varying menagerie throughout the medieval and Renaissance eras, providing artists the opportunity to see lions, wolves, bears, and lynx. There are records of a giraffe and zebra in Naples at the end of the fifteenth century, and hunting cheetahs were kept by Italian courts. Albrecht Dürer recorded a visit to the beast-garden at Brussels in 1521, where he sketched a baboon, a lynx, a chamois, and lions.

4 Francis Klingender, *Animals in Art and Thought to the End of the Middle Ages* (London: Routledge & Kegan Paul, 1971), 382–4.

5 Marie Boas, *The Scientific Renaissance, 1450–1630* (New York: Harper Brothers, 1962), 52.

6 A.M. Lysaght, *The Book of Birds* (London: Phaidon, 1975), pl. 15.

7 English needlework was famous on the Continent, and animal drawings were popular for embroidery. Witness the toucan embroidered by Mary Queen of Scots in 1570, probably based on Konrad Gesner's illustration (copied from Thevet), and labelled 'Byrd of America': illustration in Brian Ford, *Images of Science: A History of Scientific Illustration* (London: The British Library, 1992), 58.

8 Otto Pacht, 'Early Italian Nature Studies and the Early Calendar Landscape,' *Journal of the Warburg and Courtauld Institutes* 1/13 (1950): 17.

9 Klingender, *Animals in Art*, 483.

10 William M. Conway, ed., *The Writings of Albrecht Dürer* (London: Peter Owen, 1958), 247. Martyn Rix quotes the same passage in more modern language: 'study nature diligently. Be guided by nature and do not depart from it, thinking that you can do better yourself. You will be misguided for truly art is hidden

in nature, and he who can draw it out possesses it': *The Art of the Botanist* (Guildford: Lutterworth, 1981), 27.

11 Svetlana Alpers, *The Art of Describing: Dutch Art in the 17th Century (*Chicago: University of Chicago Press, 1983), 158–9.

12 Fritz Korenyi notes in his catalogue entry on this watercolour that it 'has long been admired as one of the artist's undisputed masterpieces,' and 'one of the outstanding animal studies of the Renaissance': Jay Levenson, ed., *Circa 1492: Art in the Age of Exploration*, Cat. no. 204 (Washington, DC: National Gallery of Art; New Haven, CT, and London: Yale University Press, 1991), 298.

13 Conway, ed., *The Writings of Albrecht Dürer*, 247

14 *Mirabilis jalapa* is commonly referred to as the 'Four-o'Clock,' since the flowers open late in the afternoon. The 'Marvel of Peru' was also known to apothecaries for the purgative qualities of its tuberous root. The highly ornamental Bougainvillea is also a member of the family.

15 Quoted in William M. Ivins, Jr, *Prints and Visual Communication* (Cambridge, MA: MIT Press, 1985), 34–6.

16 *Peregrinationes in Montem Syon* was published in 1486, and its cuts of an 'ape-man' were used by Gesner, though he refused to countenance the salamander illustration and noted that the '*Salamandrae figura falsa*': Donald Percy Bliss, *A History of Wood Engraving* (London: Spring 1928 [1964]), 42.

17 Noted in Lysaght, *The Book of Birds*, 19.

18 Quoted in Paul Hulton, *America 1585: The Complete Drawings of John White* (Raleigh: University of North Carolina Press, 1984), 9.

19 Ibid., 12.

20 Ibid., 11.

21 Walter E. Houghton, Jr, 'The English Virtuoso in the Seventeenth Century (II),' *Journal of the History of Ideas* 3/2 (1942): 206.

22 Lynn Thorndike, *A History of Magic and Experimental Science*, vol. 6: *The Sixteenth Century* (New York: Columbia University Press, 1961), 267.

23 Martin Rudwick, *The Meaning of Fossils: Episodes in the History of Palaeontology* (London: Macdonald, 1972), 6.

24 Quoted in C.E. Raven, *English Naturalists from Neckham to Ray: A Study of the Making of the Modern World* (Cambridge: Cambridge University Press, 1947), 82. Turner had spent some time studying with Luca Ghini (*ca* 1490–1556), botanist at Bologna, who maintained both a herbarium and a collection of drawings of plants: noted in Edward Lee Greene, *Landmarks of Botanical History*, Part II, ed. Frank N. Edgerton (Stanford, CA: Stanford University Press, 1983), 708–9.

25 Edward Topsell, *History of Four-Footed Beasts and Serpents and Insects*, vol 1: *The History of Four-footed Beasts* (1658; repr. New York: Da Capo, 1967), 585.

26 Ibid., 15.

27 Quoted in Raven, *English Naturalists*, 155 and note.

28 There is an illustration in Hugh Honour, *The New Golden Land: European Images of America from the Discoveries to the Present Time* (New York: Pantheon, 1975), fig. 33. Gesner prepared some fifteen hundred drawings for a planned *Historia Stirpium*. After his death, the illustrations were sold, and some lost. Those that remain were rediscovered in 1929 in the University of Erlangen: Greene, *Landmarks*, 758.

29 Edward Topsell, *The Fowles of Heaven or History of Birdes*, ed. T.P. Harrison and F. David Hoeniger (Austin: University of Texas Press, 1972), xxxiii.

30 These images were also sought after by wealthy collectors, for whom the 'contrafetten' were often more important than the actual object. In the early 1600s, the collection of the Prince Bishop of Eichstätt, Johann Conrad von Gemmingen, was justly celebrated. The Bishop's collection included 'various sketches of animals, plants, herbs or other curiosities and works of art,' and, with the help of the apothecary Basilius Besler, the drawings of flowers were eventually worked into one of the greatest of the early florilegia, the *Hortus Eystettensis.* These sketches, the Bishop's garden in art, were much sought after by a number of princely collectors, including Wilhelm V. Wilhelm wrote to his agent Philipp Hainhofer that, 'should the Bishop make excuse that he lacks people to do this,' that is, the copying of the Bishop's 'sketches,' Wilhelm would 'come to an arrangement with you ourselves for the copying to be completed at the lowest possible cost': Nicholas Barker, *Hortus Eystettensis: The Bishop's Garden and Besler's Magnificent Book* (London: The British Library, 1994), 60–1.

31 Jean Anker, *Bird Books and Bird Art* (Copenhagen, 1938; repr. New York: Arno, 1974), 29.

32 Wilfrid Blunt, *The Art of Botanical Illustration*, 4th ed. (London: Collins, 1971), 156.

33 Ivins, Jr, *Prints and Visual Communication*, 3.

34 Elizabeth Eisenstein, *The Printing Press as an Agent of Change: Communiation and Cultural Transformations in Early Modern Europe* (Cambridge: Cambridge University Press, 1979), 269.

35 Ibid., 264.

36 Quoted in Bliss, *A History of Wood Engraving*, 2–3.

37 Plates 5 and 7 in Roger Manning, *Hunters and Poachers: A Social and Cultural History of Unlawful Hunting in England, 1485–1640* (Oxford: The Clarendon Press, 1993.)

38 Facsimile in Hulton, *America 1585*, fig. 4.

39 Bliss, *A History of Wood Engraving*, 47.

40 William Turner, *Turner on Birds ...*, ed. A.H. Evans (Cambridge: Cambridge University Press, 1903), 52–3.

41 Aristotle names this bird the Iynx, and, while the colouring is peculiar, the illustrator does show the characteristic details enumerated by Aristotle: 'two claws in front and two behind' and 'a tongue like that of a serpents.' Turner does not know of an English equivalent: Ibid., 149.

42 Campbell Dodgson in a catalogue of German and Flemish painting at the British Museum: Bliss, *A History of Wood Engraving*, 47.
43 David Bland, *A History of Book Illustration: The Illuminated Manuscript and the Printed Book*, 2d ed. (London: Faber & Faber, 1969), 101.
44 Barker, *Hortus Eystettensis*, 65
45 Bernadette Bucher, *Icon and Conquest: A Structural Analysis of the Illustrations of de Bry's* Great Voyages, trans. Basia Miller Gulati (Chicago: University of Chicago Press, 1981), 20.
46 Quoted in Jessica Rawson, ed., *Animals in Art* (London: British Museum Publications, 1977), 127–8.
47 Translated in Museum of Fine Arts, Boston, *Albrecht Duerer, Master Printmaker* (Boston: Museum of Fine Arts, 1971), 244. It is intriguing to speculate on the origins of the reported enmity between elephant and rhinoceroses. The Greek geographer Strabo mentions fights staged between elephants and rhinoceroses, a tradition repeated with this sixteenth-century beast, which, upon arrival in Lisbon, was led into the presence of a young elephant. The elephant fled. It was planned that the 'combat' would be tested once again with the same rhinoceros and the elephant Hanno at Rome, but the rhinoceros drowned *en route* and the fight never took place. The presence of a single horn seems also to have led to confusion between unicorns and rhinoceroses. In the bestiary tradition, it is the unicorn and the elephant that are mortal enemies. François-Marc Gagnon notes in a thirteenth-century French bestiary that the unicorn is a ferocious enemy of the elephant:

   Iceste betse est si osee
   Si conbatant et si hardie,
   A l'olifant porte envie,
   La plus egre beste del mont,
   De totes celes qui i sont.

   Quoted in François-Marc Gagnon and Denise Petel, *Hommes effarables et bestes sauvaiges: Images du Nouveau-Monde d'après les Voyages de Jacques Cartier* (Montreal: Les Éditions du Boréal Express, 1986), 154. Gesner, in *Historia Animalium*, notes that a rhinoceros 'is taken by the same means that the *Unicorn* is taken, for it is said by *Albertus, Isidorus* and *Alunnus*, that above all of the creatures they love Virgins, ...': Topsell, *A History of Four-footed Beasts ...*, 463.
48 L.C. Rookmaker, 'Two Collections of Rhinoceros Plates ...,' *Journal of the Society for the Bibliography of Natural History*, 9/1 (1978): 21. In 1790, however, James Bruce, author of *Travels to Discover the Source of the Nile*, criticized Dürer's drawing of what he rightly identified as the Asian rhinoceros: 'It was wonderfully ill-executed in all its parts, and was the origin of all the monstrous forms under which that animal has been painted ever since ... Several modern philosophers have made amends for this in our days; Mr. Parsons, Mr. Edwards, and the Count de Buffon, have given good figures of it from life; they have

indeed some faults, owing chiefly to preconceived prejudices and inattention ...' (quoted in E.H. Gombrich, *Art and Illusion: A Study in the Psychology of Pictorial Representation*, 2d ed. (Princeton, NJ: Princeton University Press, 1972), 82. That George Edwards's depiction of the rhinoceros suffered some faults is no surprise, since he laboured under the same constraints as had Dürer. Edwards based his picture on a sketch of a rhinoceros made by a passenger on a Dutch East India Company vessel. The rhinoceros died during the voyage to England, and Edwards was not able to draw it 'from life': (noted in A. Stuart Mason, *George Edwards: The Bedell and His Birds* (London: Royal College of Physicians, 1992), 28.

49 Ivins, Jr, *Prints and Visual Communication*, 69–70.

50 Honour, *The European Vision of America*, fig. 61.

51 Facsimile reproduced in Hulton, *America 1585*, fig. 4. A set of two superb digital facsimiles with transcriptions of the 1588 and 1590 editions of Thomas Harriot's account of Virginia and the de Bry engravings after John White, is available on the World Wide Web from the University of Virginia as: http://wsrv.clas.virginia.edu/~msk5d/harriot/main.html

52 I examined a number of facsimiles and reproductions of de Bry's work, but references to the original are to this edition, held at the National Library of Canada.

53 In plate XX, showing the Village of Secota, the original watercolour does not show particular vegetation. De Bry has added recognizable plants – sunflowers, corn in a number of stages of growth, pumpkins, and tobacco – to the fields, following the textual descriptions. See plate 29.

54 Bucher, *Icon and Conquest*, 38–9.

55 Vegetation and wildlife often appear emblematically in the cartographic bird's-eye view representations, as, for example, the grape vines and the deer shown in plate II, *The arriual of the Englishmen in Virginia*. In the second volume of the America series, dedicated to le Moyne's account of the Florida colony, plate V shows a map-like illustration of Portus Regalis and features an Indian village in a park-like setting with stags, out-of-proportion squash and grape vines, and a flock of turkeys. This coloured engraving of 1591 is very similar in style to the work of map-makers like Desceliers: reproduced in W.P. Cumming, R.A. Skelton, and D.B. Quinn, *The Discovery of North America, 1630–1776* (New York: American Heritage Press, 1971–2), pl. 179. See also plate 12.

56 Eisenstein, *The Printing Press*, 258.

57 Francis Haskell, *History and Its Images: Art and the Interpretation of the Past* (New Haven, CT: Yale University Press, 1993), 53.

Chapter 3: The Living Image

1 W.F. Ganong, 'The Identity of the Animals and Plants Mentioned by the Early Voyagers to Eastern Canada and Newfoundland,' *Transactions of the Royal*

Society of Canada, section II, 3rd series, vol. III (1910): 227. Where indicated, the identifications follow those made by Ganong in his 1908 paper. The rest have been made tentatively by the author.

2 Oliver P. Medsger, *Edible Wild Plants* (New York: Collier's, 1939 [1976]), 122.

3 Cornut gives a good rendering of the roots with the edible tubers ranged in a row: Jacques-Philippe Cornut, *Canadensium Plantarum aliarúmque nondum editarum Historia* ... (Paris, 1635; Sources of Science, no. 37. New York: Johnson Reprint Corp., 1966), Caput LXXVI: *Apios americana.*

4 Thanks to François-Marc Gagnon for the suggestion that the 'aux' may be the plural of 'ail,' or garlic. The wild garlic, *Allium canadense*, is common in the area and its bulbs were relished by the Native peoples. Medsger notes that Father Marquette and his party ate wild onions, probably *Allium canadense*, as their chief food in their journey from Green Bay, Wisconsin, to Chicago in 1674: Medsger, *Edible Wild Plants*, 176.

5 The wild ginger (*Asarum canadense*) is widespread along the shores of Georgian Bay, where Champlain travelled.

6 Cornut's artist has never been identified, but it is likely he was attached to the royal gardens in Paris. The practice of making watercolour drawings or oils of plants appears to have been common. John Evelyn recorded in his diary for 1 April 1644 that he visited the garden of Monsieur Morine (Morin, with whom Cornut was acquainted) and observed that the famous nurseryman had caused his best flowers 'to be painted in miniature by rare hands, and some in oil': John Evelyn, *The Diary of John Evelyn*, ed. William Bray (London, J.M. Dent, [1907]), vol. 1, 67.

7 Stannard, in introduction to reprint of Cornut, *Canadensium Plantarum,* xv.

8 Even Pliny finds much to complain about in the practice of doctors in classical times: 'Doctors learn by exposing us to risks, and conduct experiments at the expense of our lives. Only a doctor can kill a man with impunity': Pliny the Elder, *Natural History: A Selection* (London: Penguin, 1991), 265.

9 Karen Reeds, 'Renaissance Humanism and Botany,' *Annals of Science* 33 (1976): 526.

10 Ibid., 528. Pliny complains that the practice of medicine had greatly deteriorated from what it had been in Hippocrates' time: 'For people find it more agreeable to sit listening in lecture theatres than to go out into lonely places searching for different plants at the appropriate season': Pliny, *Natural History: A Selection*, 245.)

11 Brian J. Ford, *Images of Science: A History of Scientific Illustration* (London: The British Library, 1992), 99n.

12 Marie Boas, *The Scientific Renaissance,1450–1630* (New York: Harper Brothers, 1962), 52.

13 Agnes Arber, *Herbals, Their Origin and Evolution: A Chapter in the History of Botany, 1470–1670*, 2d ed. (Darien, CT: Hafner, 1970), 193.

14 Pliny the Elder, *Natural History*, ed. W.H.S. Jones (Cambridge, MA: Harvard

University Press; London: William Heinemann, 1956), 141 (Book XXV, III.6 – IV.8.). Wilfrid Blunt quotes the 1601 translation by Philemon Holland in *The Art of Botanical Illustration*, 4th ed. (London: Collins, 1971), 2:

> But what certaintie could be therein? pictures (you know) are deceitful; also, in representing such a number of colours and especially expressing the lively hew of hearbs according to their nature as they grow, no marveile if they limned and drew them out, did faile and degenerate from the first pattern and originall. Besides, they come far short of the marke, setting out hearbes as they did at one onely season (to wit, either in their floure, or in seed time) for they chaunge and alter their forme and shape everie quarter of the yeere ...

15 Quoted in Arber, *Herbals*, 201.
16 Reeds, 'Renaissance Humanism,' 531n.
17 Charles Singer, *From Magic to Science: Essays on the Scientific Twilight* (New York: Dover, 1958), 197.
18 Quoted in Reeds, 'Renaissance Humanism,' 529n.
19 Quoted in Blunt, *Botanical Illustration*, 51.
20 Arber, *Herbals,* 232–3.
21 Ibid., 246.
22 Quoted in ibid., 24–6.
23 Harold McGee, *On Food and Cooking: The Science and Lore of the Kitchen* (New York: Collier's, 1984), 212, 241.
24 There was a long tradition, dating back to classical times, of the strange and exotic arriving vaguely from the East. The idea that new plants or animals came from Turkey fit comfortably with classical assumptions, which only changed as the trickle from the western hemisphere became a flood. Gerard still shows Turkey wheat in 1636, but the New World origin of the potato and Jerusalem artichoke are noted in John Parkinson, *Paradisi in Sole, Paradisus Terrestris* (1629).
25 In Jacques Cartier, *The Voyages of Jacques Cartier*, ed. Ramsay Cook (Toronto: University of Toronto Press, 1993), 127.
26 Thevet, 'Grand Insulaire,' in *André Thevet's North America. A Sixteenth-Century View*, ed. and trans. Robert Schlesinger and A.P. Stabler (Kingston: McGill-Queen's University Press, 1986), 89–90.
27 In Cartier, *Voyages,* 80.
28 James S. Pringle, 'How "Canadian" Is Cornut's Canadensium Plantarum Historia? A Phytogeographic and Historical Analysis,' *Canadian Horticultural History* 1/4 (1988): 196. John Tradescant bought two arbor-vitae trees at Harlem in 1611 for a shilling apiece, a small price for what had once been a great rarity.
29 John Gerard, *The Herbal or General History of Plants* (London, 1633; repr. New York: Dover, 1975), 1621.
30 Nicolas Monardes, *Joyfull newes out of the newe founde worlde* ... Englished by John Frampton, Merchant. Introduction by Stephen Gasellee (New York: AMS, 1967), vol. 1, preface, n.p.

31 Ibid., 99–100. Thomas Harriot also notes Monardus on the virtues of Sassafras: 'Sassafras, called by the inhabitantes Winauk, a kinde of wood of most pleasand and sweete smel; and of most rare vertues in phisick for the cure of many diseases. It is found by experience to bee farre better and of more uses then the wood which is called Guaiacum, or Lignum vitæ. For the description, the manner of vsing and the manifolde vertues thereof, I referre you to the booke of Monardus, translated and entituled in English, The ioyfull newes from the West Indies': from the transcription of Harriot (1590) provided by the University of Virginia at: http://wsrv.clas.virginia.edu/~msk5d/hariot/main.html.

32 Gerard, *Herbal*, 859–60.

33 I have taken the quotations from the 1633 edition, which has additions and amendments by Thomas Johnson and, according to the editor of the 1975 reprint, 'represents in both text and pictures, a conspectus and summation of the finest research of its age.'

34 Paul B. Hulton, *America 1585: The Complete Drawings of John White* (Raleigh: University of North Carolina Press, 1984), pl. 50.

35 Gerard, *Herbal*, 898–9.

36 Ibid., 1179.

37 Ibid., 61. 'Dries' might be a misspelling of 'De Vries.'

38 Ibid., 860–1.

39 Ibid., 926.

40 Ibid., 1543.

41 Ibid., preface, n.p.

42 John Prest, *The Garden of Eden: The Botanic Garden and the Re-Creation of Paradise* (New Haven, CT, and London: Yale University Press, 1981), 38.

43 Quelques louanges non pareilles
Qu'ayt Appelle encore aujourd'huy,
Cet ouvrage plein de merveilles
Met Rabel au-dessus de luy.
L'Art y surmonte la Nature
Et si mon Jugement n'est vain,
Flore lui conduisoit la main
Quand il faisoit cette peinture ...

Quoted in Blunt, *Botanical Illustration*, 107.

44 See the discussion by Margery Corbett, 'The Engraved Title-Page to John Gerarde's *Herball* or *General Historie of Plantes*, 1597,' *Journal of the Society for the Bibliography of Natural History* 8/3 (1977): 223–30.

45 Blunt, *Botanical Illustration*, 100–1.

46 Corbett, 'The engraved title-page ...,' 226. De Passe may have been sensitive to correct colouring, because it has been suggested by Hulton that he worked from the watercolours of an eminent plant painter and colourist, Jacques le Moyne de Morgues. Little of le Moyne de Morgues' original North American material remains, but de Bry's engravings reveal that he must have included carefully

rendered plants. De Bry's engravers often add generic plants in the foreground of the engraving, but in at least one plate in volume 2 of *Les grands voyages* (XLI), the background vegetation captures something of North American habitat, featuring as it does cattails and perhaps a milkweed (see plate 13). In England, however, le Moyne had a chance to display his skills as a flower painter, when he prepared drawings for a book entitled *La Clef des Champs* (1586). *La Clef* was conceived primarily as a pattern-book for jewellery, painting, embroidery, tapestry, and all kinds of needlework. Le Moyne draws on many earlier 'patterns' for the animal and bird figures, but the flowers are engraved after his own drawings, which are extraordinarily lifelike and delicately coloured.

47 Arber, *Herbals*, 207. Arber also notes that Plantin employed 'certain women illuminators to colour by hand the botanical books which he produced' (p. 215).

48 Quoted in Blunt, *Botanical Illustration*, 112.

49 Crispin de Pass, *Hortus Floridus* (Utrecht: Salomon de Roy, 1615; London: Crosset, 1929), vol. 1, pl. XXXII.

50 Pringle, 'How "Canadian" ...,' 205–6.

51 It should be noted that Cumming et al., in *The Exploration of North America, 1630–1776* (Toronto: McClelland & Stewart, 1974), suggest that Cornut's drawings were made from dried specimens sent from Quebec. This assertion is contradicted by Jacques Rousseau ('Quelques jalons dans l'histoire de la botanique de la Nouvelle France, de Cartier à la fin de régime français,' in *Les botanistes français en Amérique du nord avant 1850*, ed. Jean Leroy [Paris: Centre National de la Recherche Scientifique, 1958], 151), and the fact that Cornut frequently refers to the horticulture of plants would suggest that he could observe them living, and that at least some of them were drawn from life.

52 De Charlevoix 'utilise, ordonne, souvent récrit ...': P. Fournier, *Les voyageurs naturalistes du clergé français avant la Révolution* (Paris: Paul Lechevalier & Fils, 1932), 50.

53 The letters addressed to the Duchess of Lesdiguieres are a literary device and were not written on the spot and sent at the time of writing, but worked up from notes for publication: P.-F.-X. de Charlevoix, *Journal of a Voyage to North-America Undertaken by Order of the French King containing The Geographical Description and Natural History of that Country, particularly Canada ... in two volumes* (London: R. and J. Dodsley, 1761; Ann Arbor: University Microfilms, 1966), 251.

54 'Dans la figure que j'en donne ici, j'ai représenté les feuilles comme Bauhin et Catesby les ont fait graver': P.-F.-X. de Charlevoix, *Historie et description générale de la Nouvelle France* (Paris, 1744), pl. LXIIII.

55 Svetlana Alpers, *The Art of Describing: Dutch Art in the Seventeenth Century* (Chicago: University of Chicago Press, 1993), 80.

56 The doctrine of signatures suggested that the curative properties of each plant were revealed through some outward mark or signature so that knowledgeable

people could recognize God's provision of natural remedies. The walnut, for example, was assumed to be good for ailments of the brain, since the kernel resembled the brain itself: 'the Kernel hath the very figure of the Brain, and therefore it is very profitable for the Brain' (from Coles, *Adam in Eden: or Nature's Paradise* (1657), quoted in Prest, *The Garden of Eden*, 63).

57 Alpers, *Art of Describing*, 81.

Chapter 4: The Redefinition of Landscape

1 For the complex publishing history, see N.-E. Dionne, *Hennepin: Ses voyages et ses oeuvres*, 2d ed. (Montreal: Les Editions Canadiana, n.d.), 19–40. I have chosen to quote primarily from the English translation published in 1698 and available on microfiche, but have examined in detail the original 1704 edition published at Leiden by Pierre Vander Aa, and entitled *Voyage curieux du RP Louis Hennepin, Missionaire Recollect, & Notaire Apostolique, qui contient Une Nouvelle Decouverte d'un Tres Grand Pays, Situé dans l'Amerique* ... This edition also contains a notice to the reader and I have also used some excerpts from it.

2 Samuel de Champlain, *The Works of Samuel de Champlain*, ed. H.P. Biggar, vol. 2: 1608–1613 (Toronto: The Champlain Society, 1925), 184–5.

3 Louis Hennepin, *A New Discovery of a* Vast Country in *America ... With a Continuation: Giving an Account of Attempts of the Sieur* De la Salle *upon the Mines of* St. Barbe, *&c. ... By* L. Hennepin, *now Resident in* Holland. *To which is added, Several* New Discoveries *in* North-America, *not publish'd in the* French *Edition* (London: M. Bentley, J. Tonson, H. Bonwick, T. Goodwin, and S. Manship, 1698), preface, n.p.

4 'pour donner une Idée nette de certaines choses qui se comprennent mieux, quand on en a quelque représentation devant les yeux': from Louis Hennepin, *Voyage curieux du RP Louis Hennepin, Missionaire Recollect, & Notaire Apostolique, qui contient Une Nouvelle Decouverte d'un Tres Grand Pays, Situé dans l'Amerique ...* (Leide: Chez Pierre Van der An, 1704), 'Avis au Lecteur,' n.p. (my translation).

5 'sur tout une description du Grand Saut de Niagara, qui est la plus belle & tout ensemble la plus effroyable Cascade, qui soit dans tout l'Univers': ibid. (my translation).

6 'Je vous protests ici devant Dieu, que me Rélation est fidéle & sincere, & que vous pouvez ajoûter foi à tout ce qui y est rapporté': ibid. (my translation).

7 Hennepin, *A New Discovery*, 29–30.

8 Ibid., 267–8.

9 Kalm insisted that, in Canada, Hennepin was known by the name 'The great Liar.' In a translation of a letter written originally in Swedish, Kalm states the falls are 135 French feet high, or about $147\frac{690}{1000}$ Swedish feet! In Peter Kalm, *Peter Kalm's Travels in North America: The English Version of 1770*, ed. Adolph B. Benson (New York: Dover, 1964), vol. 2, 697.

10 Hennepin, *A New Discovery*, 267.

11 Jeremy Elwell Adamson, *Niagara: Two Centuries of Changing Attitudes, 1697–1901* (Washington, DC: Corcoran Gallery of Art, 1985), 18. For an excellent analysis of images of Niagara, see also Christopher W. Lane, *Impressions of Niagara: The Charles Rand Penney Collection of Prints of Niagara Falls and the Niagara River from the Sixteenth to the Early Twentieth Century* (Philadelphia: The Philadelphia Print Shop, 1993).

12 Edward Dahl, 'The Original Beaver Map – De Fer's 1698 Wall Map of America,' *The Map Collector* 29 (December 1984): 24. The myths surrounding the beavers' industry and their modes of organization and work, shown in detail in the Guérard vignette, persist into the eighteenth century. Dahl suggests they began with Gabriel Sagard's 1632 account, though they would appear to date back even further. The idea of the beaver as the master craftsman of the animal kingdom is found in the works of the twelfth-century writer Gerald of Wales, who describes beavers using their tails like sledges. Sagard's description is certainly repeated in Nicolas Denys and in de Charlevoix's account of their habits. Baron de Lahontan had hunted the beaver with his Native companions, and in the preface to the 1703 English edition of his *New Voyages to North-America*, he suggests that those who 'doubt of the Instinct and wonderful capacity of Beavers, ... need only cast their Eyes upon the Great Map of *America*, drawn by Sieur *de Fer*, and grav'd at *Paris* in the year 1698.' A later and more scientific observer, Samuel Hearne (1795), heaped scorn on these 'fictions.'

13 2 Kings 2: verse 11 (King James Version): 'And it came to pass, as they still went on and talked, that, behold, there appeared a chariot of fire and horses of fire, and parted them both asunder; and Elijah went up by a whirlwind into heaven.'

14 Kenneth Clark, *Landscape into Art* (London: John Murray, 1949; new ed., 1976), viii.

15 John Walsh, 'Skies and Reality in Dutch Landscape,' in *Art in History, History in Art: Studies in Seventeenth-Century Dutch Culture*, ed. David Freedberg and Jan de Vries (Santa Monica, CA: The Getty Center for the History of Art and the Humanities, 1991), 106.

16 Clark, *Landscape into Art*, 59.

17 William C. Sturtevant, 'The Sources of European Imagery of Native Americans,' in *New World of Wonders: European Images of the Americas, 1492–1700*, ed. Rachel Doggett, (Washington, DC: Folger Shakespeare Library, 1992), 32: 'Although accuracy was certainly a goal of artists in the sixteenth and seventeenth centuries, it was very often subordinated to stylistic and compositional considerations and was influenced by efforts to address the expectations of the audience.'

18 Wayne Franklin, *Discoverers, Explorers, Settlers: The Diligent Writers of Early America* (Chicago: University of Chicago Press, 1979), note to plate 1.

19 William Wood, 'Excerpts from *New England's Prospect*,' in *This Incomperable*

Land: A Book of American Nature Writing, ed. Thomas J. Lyon (New York: Penguin, 1989), 98.

20 Howard Mumford Jones, *O Strange New World. American Culture: The Formative Years* (New York: Viking, 1964), 28–33.

21 Note Xenophon, *On Hunting* (*Cynegiticus*); Frederick II, *De Arte Venandi cum Avibus*; Gaston Phébus, *Livre de chasse*; Tuberuile, *The Booke of Faulconrie or Hauking* ...; and Cockayne, *A Short Treatise of Hunting*.

22 Keith Thomas, *Man and the Natural World: Changing Attitudes in England, 1500–1800* (Harmondsworth: Penguin, 1984), 145.

23 Sir Thomas Cockayne, *A Short Treatise of Hunting, 1591*, with an introduction by W.R. Halliday (Oxford: Humphrey Milford, Oxford University Press, 1932).

24 Franklin, *Discoverers, Explorers, Settlers*, 23–4.

25 Samuel Eliot Morison, *The European Discovery of America: The Northern Voyages, AD 500–1600* (New York: Oxford University Press, 1971), 594. Davis, in the account of his first voyage in 1585, describes a polar bear hunt: 'but when wee came neere the shore, wee found them to be white beares of a monstruous bignesse: we being desirous of fresh victual and the sport, began to assault them ...': quoted in W.P. Cumming, R.A. Skelton, and D.B. Quinn, *The Discovery of North America, 1630–1776* (New York: American Heritage Press, 1971–2), 235.

26 'Of the Beasts that Live on the Land,' in Wood, 'Excerpts,' 98.

27 'Of the Birds and Fowls Both of Land and Water,' ibid., 103.

28 Champlain, *Les voyages de Samuel Champlain, Saintongeais, Père du Canada*, ed. Hubert Deschamps (Paris: Presses Universitaires de France, 1951), Premier voyage, 1604, 13.

29 Lescarbot's description of such natural plenty is similar to the image portrayed by White, and it would be difficult to guess from this rhapsody on the abundant fish and game that scurvy ravaged the settlement in the winter.

30 I have quoted from the 1703 English edition, *New Voyages to North-America. Written in* French *by the Baron Lahontan, Done into* English ... 2 vols. (London: H. Benwicke and others, 1703; repr. New York: Reuben Gold Thwaites, Burt Franklin, 1905 [1970]). Some quotations are taken, however, from the 1705 French edition published in Amsterdam, which includes a different preface.

31 'un Gentilhomme curieux & de bon sense ... Jeune & plein de feu ... la fatigue & le peril ne le rebutoient point ... Pendant ces voyages il tenoit regître de tout ce qui est à la portée d'u Cavalier d'esprit ...': Baron de Lahontan, *Voyages du Baron La Hontan dans l'Amérique septentrionale, tome premier* (Amsterdam: Chez François l'Honoré 1705; Montréal: Éditions Elysée, 1974), preface (my translation).

32 The 'Elk' is in French the 'orignal,' more properly translated as 'moose.' De Lahontan, *New Voyages to North-America* ..., vol. 1, 108. In the 1704 French edition, he describes his correspondent as 'un grand exterminateur de gros & de petits pieds': *Voyages*, vol. 1, 93.

33 H.P. Biggar, introduction to Marc Lescarbot, *The History of New France*, (Toronto: The Champlain Society, 1914), vol. 1, xi.
34 Champlain, *Works*, vol. 2, 176–7. Place Royale is the present location of Point à Callière in Montreal.
35 Nicolas Denys, *The Description and Natural History of the Coast of North America (Acadia)*, trans. and ed. W.F. Ganong (Toronto: The Champlain Society, 1908), 351–3.
36 Ibid., 359.
37 'Of the Birds and Fowls Both of Land and Water, ' in Wood, 'Excerpts,' 105.
38 Gabriel Sagard, *The Long Journey to the Country of the Hurons*, ed. G.M. Wrong; trans. H.H. Langton (Toronto: The Champlain Society, 1939), 38.
39 Ibid., 47.
40 Thomas calculates that half the population at the end of the seventeenth century ate more than 147 pounds of meat per person per annum: Thomas, *Man and the Natural World*, 26.
41 De Lahontan, *New Voyages to North-America*, vol. 1, 355. In the 1704 French edition, de Lahontan recounts, not that he made pancakes, but that 'on est obligé d'y mettre de l'eau pour en faire des omelettes': *Voyages*, vol. 2, 51.
42 P.-F.-X. de Charlevoix, *Journal of a Voyage to North-America Undertaken by Order of the French King containing The Geographical Description and Natural History of that Country, particularly Canada ... in two volumes* (London: R. and J. Dodsley, 1761; Ann Arbor: University Microfilms, 1966), 239.
43 Thomas, *Man and the Natural World*, 200–1.
44 De Lahontan, *New Voyages to North-America*, vol. 1, 421. In the 1704 French edition, 'Fuzee' is 'fusil,' or gun. De Lahontan, *Voyages*, vol. 2, 101.
45 Plate in volume 1 of Michel Devèze, *La vie de la forêt française au XVI*[e] *siècle* (Paris: École Pratique des Hautes Études, VIe section; SEVPEN, 1961).
46 George Turberuile, *The Booke of Faulconrie or Hauking, for the onely delight and pleasure of all Noblemen and Gentlemen, London 1575* (The English Experience, no. 93. Amsterdam: Theatrum orbis terrarum; New York: Da Capo, 1969), 112. Also see plate 10 in A.M. Lysaght, *The Book of Birds* (London: Phaidon, 1975).
47 See Edward Topsell, *History of Four-Footed Beasts and Serpents and Insects*, vol. 1: *The History of Four-footed Beasts* (1658; repr. New York: Da Capo, 1967), 96.
48 Gaston Phébus, *Livre de Chasse*, édité avec introduction ... par Gunnar Tilander (Karlshamn: E.G. Johanssons Boktryckeri, 1971), 257 and 261.
49 De Lahontan, *New Voyages to North-America*, vol. 1, preface, 9–10.
50 One wonders if beavers would have been considered such wonders of the world were it not for their ability to manage the watercourses and create meadows.
51 Thomas, *Man and the Natural World*, 194.
52 Champlain, *Works*, vol. 2, 176.
53 Ibid., 91
54 Ibid., 260.
55 Sagard, *The Long Journey*, 90.

56 Champlain, *Works*, vol. 2, 262–3.

57 Ibid., 274. Champlain was not the first explorer to complain bitterly about the mosquitoes and flies that filled the Canadian summer. Dionysus Settle, in his account of Franklin's voyage in 1577, found occasion to remark on 'certain stinging Gnattes, which bites so fiercely, that the place where they bite, shortly after swelleth and itcheth very sore' (in Dionysus Settle, *Laste Voyage into the West and Northwest Regions* [London, 1577; repr. 1969], n.p.). John Davis noted on his voyage of 1585–7 that 'a flie which is called Atuskyte ... did sting grievously' (in Morison, *The European Discovery*, 594). Jens Munk, who sailed to Hudson Bay in 1619–20, also remarked on the 'quantity of gnats ... unbearable in calm weather' (in W.A. Kenyon, ed., *The Journal of Jens Munk, 1618–1620* [Toronto: Royal Ontario Museum, 1980], 36). Even Baron de Lahontan, who hunted so assiduously, found that 'the *Maringouins*, which we call *Midges*, are unsufferably troublesome ...' (De Lahontan, *New Voyages to North-America*, vol. 1, 68). Finally, Gabriel Sagard describes the 'piece of thin stuff' he wore over his face for protection from the mosquitoes, 'since these fierce creatures would have blinded me many times, as I had been warned,' (Sagard, *The Long Journey*, 83).

58 Sagard, *The Long Journey*, 83. About one-and-a-half hours' drive north of Ottawa, there exists a small patch of old-growth forest (Shaw Woods) that may give some impression of what Champlain and others endured. The trees – hemlock, grey birch, pines – rise up to form a canopy some thirty to thirty-five metres overhead. The understory is crammed with small maples, junipers, saplings, and ferns. Deadfalls impede progress along the trails, and huge rocks loom out of the ground. The wood is dim, walking is made difficult not only by insects, but by the thin branches of saplings whipping at hands, feet, and face, even along the rudimentary trails.

59 From the *Jesuit Relations*, vol. 1, quoted in Jones, *O Strange New World*, 56.

60 François Du Creux, *The History of Canada or New France*, ed. James B. Conacher: trans. Percy J. Robinson (Toronto: The Champlain Society, 1951), 15.

61 Thomas, *Man and the Natural World*, 207–8.

62 Ibid., 255.

63 Francis Bacon, 'Of Gardens,' in *The Essays*, ed. John Pitcher (Harmondsworth: Penguin, 1987), 199–201.

64 Du Creux, *The History of Canada*, 74–5

65 Ibid., 80. Of the five predatory birds, one is shown dabbling like a duck, one grasps a large fish with beak and talon, another eats a fish on the shore (the different feet are clearly seen), while two float serenely in profile.

66 A manuscript drawing from Gerald of Wales's *Topograhic Hibernica* (1188) shows a bird with one webbed and one taloned foot diving into the water at some fish: Brian Ford, *Images of Science: A History of Scientific Illustration* (London: The British Library, 1992), 61.

67 'There is also in the foresaide Island Hispaniola, a strainge birde ... called

Amphibion, that lyveth both on the land and upon the water. His right foote is like the talant of an Eagle and his left foote like a gooses, upon this hee standeth when hee is beneath upon the land; and seinge a fishe hee flieth upon the water stayinge himselfe from sinkinge by his broade foote, but with his other crooked clawes taketh the fishe and eateth it upon some rocke, or neere adjacent tree ...': Topsell, *The Fowles of Heaven*, 118.
68 Du Creux, *The History of Canada*, 72.
69 Hennepin, *A New Discovery*, 115.
70 De Lahontan, *New Voyages to North-America*, vol. 1, 370.
71 Ibid., vol. 1, 319.
72 De Charlevoix, *Journal of a Voyage*, vol. 1, 245.
73 Settle, *Laste Voyage* , n.p.

Chapter 5: The Classification of the Visible: Part One

1 'I dedicate this piece to GOD, in humble gratitude for all the good things I have received from him in this World ...': in the preface to George Edwards, *A Natural History of Birds* (1743), part 1 of *A Natural History of Uncommon Birds and Some Other Rare and Undescribed Animals, Quadrupeds, Reptiles, Fishes, Insects, etc. In Seven Parts* ... 4 vols. (includes *Gleanings of Natural History*) (London: J. Robson, 1805–6).
2 Edwards refers to his plates as engravings, but they are, in fact, etchings, or perhaps a combination of both techniques, and he provides instructions of his technique in part IV: 'Some brief Instructions for ETCHING or ENGRAVING on *Copper Plates* with *Aqua Fortis*.'
3 Edwards appears to have been fond of parrots. He kept a 'Little Green Parrot,' which he had acquired from a London dealer. He described it as 'a brisk, lively, nimble bird, and talkative in a language unknown to me': quoted in A. Stuart Mason, *George Edwards: The Bedell and His Birds* (London: Royal College of Physicians, 1992), 32.
4 *Some Memoirs of the Life and Works of George Edwards, Fellow of the Royal and Antiquarian Societies* (London, Printed for J. Robson, Bookseller, 1776), 16n.
5 Bernard Smith, *European Vision and the South Pacific*, 2d ed. (New Haven, CT, and London: Yale University Press, 1988), 339.
6 Wilfrid Blunt, *The Compleat Naturalist: A Life of Linnaeus* (London: Collins, 1971), 120.
7 A.M. Lysaght, *Joseph Banks in Newfoundland & Labrador, 1766: His Diary, Manuscripts and Collections* (Berkeley: University of California Press, 1971), 80.
8 J.E. Smith, 'Introductory discourse on the rise and progress of Natural History,' *Transactions of the Linnaean Society*, (1791): 1–55.
9 Edwards, *A Natural History*, part I, xiv.
10 Ray and Willughby did include uncoloured illustrations in the *Ornithologiae*

Libri Tres of 1676, and engravings were featured in the work of some seventeenth-century Continental naturalists.

11 Konrad Gesner in *Historia Animalium, Libri IIII* (1558), in Alyne Wheeler, P.S. Davis, and E. Lazenby, 'William Turner's (*ca* 1508–1568) Notes on Fishes in His Letter to Conrad Gesner,' *Archives of Natural History* 13/3 (1986): 298.

12 *Some Memoirs of the Life and Works of George Edwards* ..., 16n.

13 Quoted in Joseph Ewan, *William Bartram: Botanical and Zoological Drawings, 1756–1788* (Philadelphia: American Philosophical Society, 1968), 6.

14 Letter, December 1772, in Lysaght, *Joseph Banks*, 255.

15 Edwards, *A Natural History*, part I, xiv–xv.

16 Ibid., xi.

17 *Historia Naturalis Brasiliae*, published in 1648. David Freedberg notes that 'in many copies they are coloured with extraordinary exquisiteness and accuracy ...': Freedberg, 'Science, Commerce and Art: Neglected Topics at the Junction of History and Art History,' in *Art in History, History in Art: Studies in Seventeenth-Century Dutch Culture*, ed. David Freedberg and Jan de Vries (Santa Monica, CA: The Getty Center for the History of Art and the Humanities, 1991), 393.

18 Quoted in Sharon Valiant, 'Questioning the Caterpillar,' *Natural History* 101/12 (December 1992): 50.

19 From the preface to *Metamorphosis*, quoted in Freedberg, 'Science, Commerce and Art,' 379. James Edward Smith, writing in the inaugural voulme of *Transactions of the Linnaean Society* in 1791, described Madam Merian's 'excellent work on the Surinam Insects, one of the most splendid in natural history, ... a monument of female perseverance and enthusiasm.'

20 From the preface to *Metamorphosis*, quoted in Freedberg, 'Science, Commerce and Art,' 379–80.

21 Ibid., 379.

22 'Merian's figures are slightly copied in the annexed plate': in a reprint of 'The Frog-Fish of Surinam' from the *Philosophical Transactions*, included in *A Natural History of Uncommon Birds* ..., 30.

23 Quoted in Richard Altick, *The Shows of London: A Panoramic History of Exhibitions, 1600–1862* (Cambridge, MA: The Belknap Press of Harvard University, 1978), 13.

24 Quoted in notes by Joseph Ewan, in Mark Catesby, *The Natural History of Carolina, Florida and the Bahama Islands.* 1771 ed. facs. (Savannah, GA: Beehive, 1974), ix.

25 Joseph Kastner, *A Species of Eternity* (New York: Alfred A. Knopf, 1977), 17–18.

26 Quoted in Ewan, notes, in Catesby, *The Natural History of Carolina* ..., 89–90.

27 Catesby, *The Natural History of Carolina* ..., iv. Lysaght, in *The Book of Birds* (London: Phaidon 1975), pl. 46, notes that the prairie chicken is pictured next to a dodecatheon, and that, while the chicken is to be found in the grasslands, the flower grows in the woods.

28 Catesby, *The Natural History of Carolina* ... vi–vii.

29 Ibid.
30 Quoted in Wilfrid Blunt, *The Art of Botanical Illustration*, 4th ed. (London: Collins, 1971), 50.
31 Catesby, *The Natural History of Carolina* ..., vii.
32 Charles Fothergill, 'Canadian Researches Chiefly in Natural History,' Manuscript journal, 1816–21, n.p. The manuscript is in the Fisher Rare Book Room of the Robarts Library at the University of Toronto.
33 Edwards also notes, in his redrawing of the ground squirrel, that even though Catesby had provided an excellent likeness he has included his own illustration, since 'as that expensive Work will fall into but a few Hands, I hope this Figure will prove acceptable to most of my encouragers': Edwards, *A Natural History*, part IV, pl. 181. Catesby's ground squirrel, or chipmunk, is by far the more felicitous rendering, most likely owing to the fact that Catesby had seen the animal alive in its native habitat.
34 The majority of his plates are of birds from North America, Africa, India, and China, with a few unusual European species. W. McAtee notes that, in all, Edwards figures 100 North American species of birds, from a total of 394 birds and beasts: McAtee, 'North American Birds of George Edwards,' *Journal of the Society for the Bibliography of Natural History* 215 (1950): 194–205.
35 Edwards, *A Natural History*, part I, xvi–xvii.
36 Ibid., part II, 113.
37 Ibid., 114.
38 Catesby, *The Natural History of Carolina* ..., iv–vi. The very few are probably those he copied from John White.
39 Edwards, *A Natural History*, part I, pl. 7.
40 Ibid., pl. 17.
41 Ibid., part II, pl. 60.
42 Ibid., part I, pls. 2 and 12.
43 Ibid., pls. 1 and 2.
44 Ibid., part III, pl. 130.
45 Lysaght, *Joseph Banks*, 92.
46 Edwards, *A Natural History*, part I, pl. 4.
47 Catesby, *The Natural History of Carolina* ..., iv. Even for the observer on the spot, fish were notoriously hard to render. Catesby mentioned that he was forced to procure a succession of live fish, to paint one after the other, since, unlike birds, fish 'do not retain their Colours when out of their Element.'
48 Edwards, *A Natural History*, part II, pls. 103 and 104.
49 Noted by William Hunter (1718–1783). Moose were painted and drawn by George Stubbs in 1770 and 1773, and an engraved portrait of one of Stubbs's moose is included in Thomas Pennant's *Arctic Zoology* (1792). See W.D.I. Rolfe, 'William Hunter (1718–1783) on Irish "elk" and Stubb's Moose,' *Archives of Natural History* 11/2 (1983): 263.
50 Edwards, *A Natural History*, part I, pl. 22.

51 Ibid., part II, 113.
52 Ibid., part I, pl. 4.
53 Plants were, of course, the exception, since most plant collectors wished to grow their own examples of new genera, and much effort was expended on shipping viable samples as well as herbarium specimens. Procedures for collecting botanical materials were well established by the seventeenth century, and the Royal Society's 1669 recommendations to Edward Diggs, who was setting out for Virginia, describe the method: 'To send all sorts of Berries, Grasses, Grains and Herbs, growing in Virginia, and to wrapp up the Seeds very dry in paper, to send Seeds or Berrys, when they are ready to dropp off, with as much husk and skins upon ym, as may be. to wrap up Roots in Mosse or light Earth, and to keep them from any dashing of Sea-water in the voyage; to gather ye smaller fruits, and dry ym in ye Air and in the Shade, to open fruits of a larger kind and ye Stones and Kernels being taken out, to dry ym; to set Plants or young Trees in half Tubbs of Earth, arched over with hoops, and cover'd with matts, to preserve ym from the dashing of Sea-water, giving then Air every day, ye weather being fair, and watering ym with fresh water every day': quoted in Raymond P. Stearns, *Science in the British Colonies of America* (Urbana: University of Illinois Press, 1970), 697. Despite their best care, many plant shipments suffered mightily *en route*, owing to the hazards of sea passage. Joseph Banks complained that 'on the fifth of Nov[r] we had a very hard Gale of Wind of the Western Islands which has almost ruind me in the Course of it we shipp'd a Sea which Stove in our Quarter & almost Filld the Cabbin with water in an instant where it washed backward & forward with such rapidity that it Broke in peices Every chair & table in the Place among other things that Sufferd my Poor Box of Seeds was one which was intirely demolish'd as was my Box of Earth with Plants in it which Stood upon Deck': Banks's journal entry, in Lysaght, *Joseph Banks*, 151.
54 Quoted in W. McAtee, 'North American Bird Records in the "Philosophical Transactions," 1665–1800,' *Journal of the Society for the Bibliography of Natural History*, 3/1 (1953): 57.
55 '... vous seres surpris que j'aie encore suivre mon ancienne methode pour la conservation des oiseaux, et que je n'aie pas Suivre celle que vous m'avés apprise par vous derniers lettres: mais vous m'excuserés quand vous sçaurés que cette methode est impracticable dans les païs surtout ou j'ai faites mes collections on n'y trouve ni baril ni tonellier, ni alum ni Sel. on a bien de la peine a avoir le dernier article pour les besoin de la vie, et nos colonies manquent absolument de tout, point d'alum, point d'esprit de vin, point de drouges [*sic*], etc.': quoted in Paul L. Farber, 'The Development of Taxidermy and the History of Ornithology,' *ISIS*, 68/244 (1977): 552n. (my translation).
56 Quoted in Charlotte Porter, 'The Drawings of William Bartram (1739–1823), American Naturalist,' *Archives of Natural History* 16/3 (1989): 291.
57 Kastner, *Species of Eternity*, 49.

58 From Banks's journal entry for 2 and 3 August 1762, in Lysaght, *Joseph Banks*, 129.
59 Farber, 'The Development of Taxidermy,' 552n.
60 Letter from Peter Collinson to William Bartram, 24 April 1742, in Catesby, *The Natural History of Carolina* ..., note to plate 18.
61 Edwards, *A Natural History*, part I, pl. 6.
62 Letter from Collinson to Bartram, 26 January 1738/9, quoted in Catesby, *The Natural History of Carolina* ..., 98.
63 Edwards, *A Natural History*, part I, pl. 9.
64 Ibid., part III, pl. 107.
65 Ibid., part II, pl. 100.
66 Ibid., pl. 64.
67 Ibid., pl. 73.
68 Ibid., part I, 33. Birds often arrived from the Orient minus heads and feet (likely to prevent deterioration), and in the sixteenth century many theories were advanced as to why the bird of paradise was legless. Edwards refers a number of times to curious drawings from China. English merchants and others living overseas in India or the Far East often recruited native painters to depict natural-history subjects in drawings and watercolours, which they sent home. Edwards also relied on drawings made by other respected natural-history painters such as Catesby and William Bartram.
69 *Some Memoirs of the Life and Works of George Edwards* ..., 10–11. Kastner notes that the American collector John Bartram would mark his shipments to Peter Collinson in England with the alternative address of Antoine de Jussieu at the Jardin du Roy in Paris, confident that, if the ship were captured, the prestige attached to the work of the naturalists would ensure delivery to an appropriate recipient: Kastner, *A Species of Eternity*, 49.
70 From *Gleanings of Natural History*, part III: quoted in Mason, *George Edwards*, 44.
71 Edwards, *A Natural History*, part IV, 213.
72 Ibid., part II, 112.
73 Ibid., part I, xix-xx.
74 Ibid., *Gleanings*, part III, x.
75 Ibid., part I, xx.
76 *Some Memoirs of the Life and Works of George Edwards* ..., 21n.
77 Edwards, *A Natural History, Gleanings*, part III, viii–ix.
78 Linnaeus to C.J. Trew, quoted in Gerta Calmann, *Ehret, Flower Painter Extraordinary: An Illustrated Life* (Oxford: Phaidon, 1977), 97.

Chapter 6: The Classification of the Visible: Part Two

1 *Some Memoirs of the Life and Works of George Edwards, Fellow of the Royal and Antiquarian Societies* (London: Printed for J. Robson, Bookseller, 1776), 9.

2 The German edition, printed at Nuremberg, had plates re-engraved by Johann Seligmann. This edition was translated into both French (1768–76) and the Dutch edition noted above. In my possession is a single plate of a kingfisher with titles in German, Latin, and French.
3 For the publication history, see Sacheverell Sitwell, H. Buchanan, and J. Fisher, *Fine Bird Books, 1700–1900* (New York: Atlantic Monthly Press, 1990), 93. Sitwell et al. call Edwards's *Natural History* (including *Gleanings*) 'one of the most important of all Bird Books' at its date of issue.
4 George Edwards, *A Natural History of Uncommon Birds and Some Other Rare and Undescribed Animals, Quadrupeds, Reptiles, Fishes, Insects, etc. In Seven Parts* ... 4 vols. (includes *Gleanings of Natural History*) (London: J. Robson, 1805–6), pt. IV, xxx.
5 Quoted in A.M. Lysaght, *Joseph Banks in Newfoundland & Labrador, 1766: His Diary, Manuscripts and Collections* (Berkeley: University of California Press, 1971), 104.
6 From Pennant's 'Literary Life,' quoted in Bernard Smith, *European Vision and the South Pacific*, 2d ed. (New Haven, CT, and London: Yale University Press, 1988), 11–12.
7 Lysaght notes that, according to Taylor White, a pair of caribou were brought to England in 1769: Lysaght, *Joseph Banks,* 167, n108.
8 Smith, *European Vision and the South Pacific*, 14.
9 W.D. Ian Rolfe, 'William Hunter (1718–1783) on Irish 'elk' and Stubbs's Moose,' *Archives of Natural History* 11/2 (1983): 265.
10 Sir Joshua Reynolds, Discourse III, 1770, 14–21, in *Discourses on Art*, ed. Robert R. Ware (New Haven, CT, and London: Yale University Press, 1975), 41.
11 Discourse III, 1770, 287–91, in ibid., 50.
12 Discourse XI, 1782, 55–63, in ibid., 192.
13 Discourse XI, 1782, 273–9, in ibid., 199.
14 Edwards, *A Natural History*, part I, xviii.
15 Rolfe, 'William Hunter,' 272–3.
16 Quoted in Frans A. Stafleu, *Linnaeus and the Linnaeans: The Spreading of Their Ideas in Systematic Botany, 1735–1789* (Utrecht: Oostnoek's Uitgenersmaatschappi H.V., 1971), 11–12.
17 It is likely that it follows the traditional order of the pharmacopoeia, given Cornut's background.
18 *Linnaeus and the Linnaeans*, 62.
19 Ibid., 51.
20 Ibid., 33.
21 W.T. Stearn, 'Notes on Linnaeus's "Genera Plantarum,"' in C. Linnaeus, *Genera Plantarum, fifth edition 1754*. Historiae naturalis classica 3. Fres. repr. (Weinheim, 1960), ix.
22 Ibid., xiii.
23 *Linnaeus and the Linnaeans*, 228.

24 C. Linnaeus, *A System of Vegetables ... Translated from the Thirteenth edition (as published by Dr. Murray) of the Systema Vegetabilium ...* (Lichfield: John Jackson for Leigh and Sotheby, London, 1783). Linnaeus's sentiments were not necessarily original, and these words echo the prescriptions of Sébastien Vaillant in *Botanicon parisiense:* 'Enfin on peut dire que la gloire de la Botanique s'augmente surtout en ce tems, ou de toutes parts des hommes habiles et nés pour de tels misteres, observent dans les lieux, ou ils demeurent, chaque plante en particulier, qu'ils les marquent, les decrivent, en donnent des desseins, et les conservent entre deux feuilles de papier, en fin que par là ils augmentent les monumens immortels d'un art, qui ne perira jamais': In *Botanicon parisiense, ou Denombrement par ordre alphabetique des Plantes Qui se trouvent aux environs de Paris* (Leiden and Amsterdam, 1727), preface, n.p.

25 *Linnaeus and the Linnaeans*, 77–8.

26 Ibid., 38.

27 Wilfrid Blunt, *The Art of Botanical Illustration*, 4th ed. (London: Collins, 1971), 156.

28 Wilfrid Blunt, *The Compleat Naturalist: A Life of Linnaeus* (London: Collins, 1971), 173. The plates are, respectively, on pages 127 and 222.

29 Ibid., 107.

30 Rolfe, 'William Hunter,' 275.

31 Ibid., 284n.

32 Ibid., 266–7. Hunter's careful observations were an attempt to understand the species, to distinguish it from European elk, to determine if moose had existed at some time in Europe. Both moose and European elk belong to the genus *Alces*, and some zoologists believe they differ only in race. Both were spread over the greater part of Europe at the end of the last Ice Age: Léon Bertin et al., *Larousse Encyclopedia of Animal Life* (London: Hamlyn, 1971), 600.

33 Rolfe, 'William Hunter,' 270.

34 Edwards, *A Natural History*, part III, x.

35 Quoted in Richard Mabey, *The Flowering of Kew: 200 Years of Flower Paintings from the Royal Botanic Gardens* (London: Century, 1988,) 27. Franz Bauer had been one of Jacquin's artists for the *Icones plantarum rariorum* (1781–93).

36 *Linnaeus and the Linnaeans*, 31.

37 Quoted in Lysaght, *Joseph Banks*, 91–2.

38 Edwards, *A Natural History*, part I, 111. See chapter 1 for a discussion of the nature of the su.

39 Here the porcupine is accompanied by several insects, and Moses Harris (1730–1788?) presumably went on to author *An Exposition of English Insects ...* (1766). An interesting note accompanies the issue, stating that at the print shop in St Martin's Lane 'may be seen the real butterflies brought from new Halifax.' Harris did not copy the drawing from Ellis, as stated by W.K. Morison in an article titled 'The Procupine [*sic*] Map' in the *ACML Bulletin*, 62/18 (March 1987). The quill detailed by Edwards and depicted on the Harris map is missing from the Ellis engraving.

40 Eleazar Albin, *A Natural History of Spiders and Other Curious Insects* (London), n.p.
41 Edwards, *A Natural History*, part III, pl. 291.
42 Stearn, 'Linnean Classification, Nomenclature and Method,' 247. Since animals have the power of locomotion, the inclusion of habitat in diagnosis made good sense.
43 Blunt, *The Compleat Naturalist*, 232.
44 Martin Kemp, *The Science of Art: Optical Themes in Western Art from Brunelleschi to Seurat* (New Haven, CT: Yale University Press, 1990), 1.
45 Stafleu, *Linnaeus and the Linnaeans*, 28.
46 Quoted in Smith, *European Vision and the South Pacific*, 11.
47 *Some Memoirs of the Life and Works of George Edwards ...*, 13–14.

Chapter 7: A Country Observed

1 Quoted in Richard Mabey, *Gilbert White: A Biography of the Author* of The Natural History of Selborne. (London: Century, 1986), 109.
2 'Directions for Sea-men, bound for far Voyages, 1662–3' in Raymond Stearns, *Science in the British Colonies of America* (Urbana: University of Illinois Press, 1970), 687.
3 Michael A. Dennis. 'Graphic Understanding: Instruments and Interpretation in Robert Hooke's *Micrographia*,' *Science in Context* 3/2 (1989): 309–64.
4 Quoted in Mabey, *Gilbert White*, 87.
5 James Isham, *James Isham's Observations on Hudson's Bay and Notes and Observations on a book entitled* A Voyage to Hudson Bay in the Dobbs Galley, 1749. (London: Hudson's Bay Record Society, 1949), n.p.
6 In a note on the swans included in chapter 10 of *Journey*, Hearne comments on the formation of the windpipe in the two types of swans that frequent Hudson Bay (281n). In the same chapter, Hearne also notes that he attempted to examine the lice of a mouse: 'At that time I had an excellent microscope, and endeavoured to examine them, and to ascertain their form, but the weather was so exceedingly cold, that the glasses became damp with the moisture of my breath before I could get a single sight': Samuel Hearne, *A Journey from Prince of Wales's Fort in Hudson's Bay to the Northern Ocean ... in the years 1769, 1770, 1771, 1772* (Dublin, 1796; repr. Toronto: Macmillan, 1958), 249.
7 In John Bartram, *Observations on the* Inhabitants, Climate, Soil, Rivers, Productions, Animals, and other matters worthy of Notice ... (London: J. Whiston and B. White, 1751), 94.
8 Ibid., 81.
9 Ibid., 83–4.
10 Ibid., 87–8.
11 Henry Popple published a twenty-sheet 'Map of the British Empire in America' in 1773, which includes an inset of Niagara. The Kalm illustration is based on this re-engraving of the Hennepin image.

12 Like the Dürer rhinoceros, the Hennepin reference image, in this case the *Gentleman's Magazine* engraving, endured for a considerable period. It is repeated again towards the end of the century by Robert Hancock (1730–1817), who produced a hand-coloured engraving with inscription in French and English (see chapter 4).
13 Quoted in Didier Prioul, 'British Landscape Artists in Quebec: From Documentary Views to a Poetic Vision,' in *Painting in Quebec, 1820–1850: Essays,* ed. Mario Béland (Quebec: Musée du Québec, 1992), 53.
14 R.H. Hubbard, ed., *Thomas Davies, c. 1737–1812* (Ottawa: National Gallery of Canada, 1972), 12.
15 William Gilpin, *Three Essays: on Picturesque Beauty; on Picturesque Travel; and on Sketching Landscape, to which is added a poem on Landscape Painting,* 2d ed. (London: R. Blamine, 1794), 42.
16 Hubbard, *Thomas Davies*, 28.
17 Ibid.
18 Ibid., 62.
19 Thomas Pennant, *Arctic Zoology* (London, 1792), vol. 2, 418: 'In a second of these, which he [Davies] had alive, he observed that it frequently set up two feathers over the eyes ... He had this bird alive some time, but could not make it tame.'
20 Hubbard, *Thomas Davies*, 18.
21 Ibid., 16.
22 It should be noted that Sandby accompanied Joseph Banks and Daniel Solander on a plant-hunting expedition to Wales in 1773: see A.M. Lysaght, *Joseph Banks in Newfoundland & Labrador, 1766: His Diary, Manuscripts and Collections* (Berkeley: University of California Press, 1971), 51.
23 Davies shows himself seated sketching before the falls of the Seneca in 1761 (plate 16 in Hubbard), and again, on the rocks, in *The Falls of Chaudière near Quebec*, executed in 1787 (plate 48 in Hubbard).
24 Quoted in Bernard Smith, *European Vision and the South Pacific*, 2d ed. (New Haven, CT, and London: Yale University Press, 1988), 12.
25 Quoted in ibid., 14.
26 Ibid., 34.
27 Thomas Pennant, *Introduction to the Arctic Zoology*, 2d ed. (London: Robert Faulder, 1792), ccxlii
28 Ibid., advertisement, n.p.
29 Andrew Graham, *Notes and Observations on Hudson's Bay, 1767–1791,* ed. Glyndur Williams (London: Hudson's Bay Record Society, 1969), 389. Glyndur Williams goes to some trouble to correct the impression fostered by both Hutchins and Pennant that Hutchins was the prime informant. The 'manuscript observations in a large folio volume' were primarily the work of Graham.
30 Quoted by Williams in ibid., 81n.
31 Pennant, *Introduction to the Arctic Zoology*, cclxxii.
32 Hearne, *Journey*, 230.

33 Ibid., 149.
34 Pennant, *Introduction to the Arctic Zoology.*
35 Graham, *Notes and Observation*, 42n. Graham did, of course, send back specimens which were figured by Pennant's artists in *Arctic Zoology.*
36 Lysaght, *Joseph Banks*, 104.
37 All animals are shown in winter coats and plumage, and there was at the time some controversy over the nature of winter colour change in northern animals and birds: see the notes in Lysaght, *Joseph Banks*, 79–83.
38 Pennant noted, in *Introduction to the Arctic Zoology*, that in one of the plates accompanying Cook's voyages 'is given the only accurate figure of walrus I have ever seen' (cclvii) The walrus was painted by John Webber, perhaps at Cook's urging, and Cook himself had noted in his journal that he 'had no were seen a good drawing of one': quoted in Bernard Smith, *The Art of Captain Cook's Voyages*, 3 vols. (Melbourne: Oxford University Press, 1985), 116.
39 Quoted in Smith, *European Vision*, 16–18.
40 Ibid., 6. Walter's concerns echo George Edwards. Banks took a copy of Walter's edition with him on his voyage to the Pacific.
41 Ibid., 9.
42 Pennant also used information from DesBarres, and the watercolour of Sable Island in the 1792 personal edition of *Arctic Zoology* (now at McGill University, Blacker-Wood Library of Biology) is similar to the DesBarres drawing.
43 Pennant, *Introduction to the Arctic Zoology,* advertisement, n.p.
44 Henry Ellis, *A Voyage to Hudson's Bay by the Dobbs Galley and California ...* (London: H. Whitridge, 1748).
45 I.S. Maclaren, 'The Aesthetics of Back's Writing and Painting from the First Overland Expedition,' in *Arctic Artist: The Journals and Paintings of George Back, Midshipman with Franklin, 1819–1822*, ed. C.S. Houston (Montreal: McGill-Queen's University Press, 1994), 298.
46 Secretary of the Admiralty to Richardson, 25 March 1819, in *Arctic Ordeal: The Journal of John Richardson, Surgeon-Naturalist with Franklin, 1820–1822,* ed. C.S. Houston (Kingston: McGill-Queen's University Press, 1984), xxiii.
47 Noted in Back's journal account on 14 September 1821: Houston, ed., *Arctic Artist*, 172.
48 Noted in Back's journal (Houston, ed., *Arctic Artist*, 117) and in Richardson's journal (Houston ed., *Arctic Ordeal*, 83).
49 Maclaren, 'Aesthetics,' 302.
50 Noted in Back's journal: Houston, ed., *Arctic Artist*, 42.
51 Ibid., 106.
52 Ibid., 138.
53 Maclaren, 'Aesthetics,' 297. Hawkesworth notes in his account of the first Cook expedition, at Tahiti in April 1769, that the flies made it almost impossible for 'Mr. Parkinson, Mr. Banks' natural-history painter, to work: for they not only covered his subject so that no part of its surface could be seen, but even eat the

colour off the paper as fast as he could lay it on': quoted in F.C. Sawyer, 'Some Natural History Drawings Made during Captain Cook's First Voyage round the World,' *Journal of the Society for the Bibliography of Natural History*, 2/5 (1949): 190.

54 All included in Sketchbook No. 1, National Archives of Canada.

55 Back's journal, in Houston, ed., *Arctic Artist*, 60.

56 Hood's journal, in C.S. Houston, ed., *To the Arctic by Canoe, 1819–1821: The Journal and Paintings of Robert Hood, Midshipman with Franklin* (Montreal: McGill-Queen's University Press, 1974), 43.

57 Ibid., 58.

58 Richardson's journal contains only a few references to the moose. The expedition carried a copy of the 1817 edition of Cuvier's *Le Règne Animal.*

59 One would be hard put to agree with Houston's estimation that 'Hood's paintings are much superior to the contemporary American bird paintings of the famous Alexander Wilson' (168). Wilson's paintings, while stiff, proceed from a profound understanding of the bird itself, and a number of them were copied by Audubon. The scope of Wilson's work (*American Ornithology*, [Philadelphia, 1814–18]) is also immeasurably greater.

60 Back's journal, in Houston, ed., *Arctic Artist*, 137.

61 Smith, *European Vision*, 4.

62 Humboldt's *Personal Narrative of Travels to the Equinoctial Regions of America during the years 1799–1804* was translated into English between 1814 and 1826.

63 Quoted in Smith, *European Vision*, 203–4. Mrs Sabine was presumably the wife of Edward Sabine (1788–1883), who made two Arctic voyages and recorded the observations from a worldwide magnetic survey, first suggested to the Royal Society by Humboldt.

64 Hood's journal, in Houston, ed., *To the Arctic by Canoe,* 93n.

65 Ibid., 116.

66 Ibid., 64. Hood's description is akin to Linnaeus's paen to a northern spring. Linnaeus begins a paper to Celsius on the sexuality of plants with these words: 'In spring, when the bright sun comes nearer to our zenith, he awakens in all bodies the life that has lain stifled during the chill winter. See how all creatures become lively and gay, who through the winter were dull and sluggish! See how every bird, all the long winter silent, bursts into song! See how all the insects come forth from their hiding-places where they have lain half dead, how all the plants push through the soil, how all the trees which in winter were dormant now break into leaf! ...': quoted in Wilfrid Blunt, *The Compleat Naturalist: A Life of Linnaeus* (London: Collins, 1971), 34.

67 Hood's journal, in Houston, ed., *To the Arctic by Canoe,* 115–16.

68 Back's journal, in Houston, ed., *Arctic Artist*, 55.

69 Maclaren, 'Aesthetics,' 294–5.

70 Richardson to Back, 9 June 1821, in Houston, ed., *Arctic Artist*, xxix.

71 Back's journal, in ibid., 19.
72 Ibid., 38.
73 Ibid., 105.
74 François Du Creux, *The History of Canada or New France*, ed. James B. Conacher, trans. Percy J. Robinson, (Toronto: The Champlain Society, 1951), 15.

Conclusion: Drawing and Nature

1 'To use human-made objects as sources for history and other social sciences, we have to "read them," a metaphor for interpreting them: from Jacques Maquet, 'Objects as Instruments, Objects as Signs,' in S. Lubar and W.D. Kingery, ed., *History from Things: Essays on Material Culture* (Washington DC: Smithsonian Institution Press, 1993), 39.
2 Keith Thomas, *Man and the Natural World: Changing Attitudes in England, 1500–1800* (Harmondsworth: Penguin, 1984), 15.
3 Ibid., 67.
4 Michael A. Dennis. 'Graphic Understanding: Instruments and Interpretation in Robert Hooke's *Micrographia*,' *Science in Context* 3/2 (1989): 335.
5 Ibid., 345.
6 Thomas Falconer, in Bernard Smith, *European Vision and the South Pacific,* 2d ed. (New Haven, CT: Yale University Press, 1988), 12.
7 Martin J.S. Rudwick, 'The Emergence of a Visual Language for Geological Science, 1760–1840,' *History of Science* 14 (1976): 150.
8 Mary G. Winkler and Albert Van Helden, 'Representing the Heavens: Galileo and Visual Astronomy' *Isis* 83 (1992): 208–9.
9 Illustrations were to be used 'so that students may more easily recognize objects that cannot be very clearly described in words' (1558): quoted in I. Bernard Cohen, *Album of Science: From Leonardo to Lavoisier, 1450–1800* (New York: Charles Scribner's Sons, 1980), 150.
10 Agricola used illustrations 'lest description which are conveyed by words should either not be understood by men of our own times, or should cause difficulty to posterity,': From *De re metallica*, which included 291 illustrations; quoted in Harry Robin, *The Scientific Image: From Cave to Computer* (New York: Harry N. Abrams, 1992), 98.
11 Martin Kemp, 'Taking It on Trust: Form and Meaning in Naturalistic Representation,' *Archives of Natural History* 17/2, (1990): 128.
12 Francis Haskell, *History and Its Images: Art and the Interpretation of the Past* (New Haven CT: Yale University Press, 1993), 79.
13 Sir Ernst Gombrich, 'Mirror and Map: Theories of Pictorial Representation,' *Philosophical Transactions of the Royal Society*, 270/903 (1974): 122.
14 Quoted in Neil Harris, *Cultural Excursions: Marketing Appetites and Cultural Tastes in Modern America* (Chicago and London: University of Chicago Press, 1993), 312.

15 Ibid., 313.
16 Ibid., 308.
17 William M. Ivins, *On the Rationalization of Sight, with an Examination of the Renaissance Texts on Perspective,* Metropolitan Museum of Art Papers no. 8 (NewYork: Metropolitan Museum of Art, 1938), 13.
18 Eugene S. Ferguson, 'The Mind's Eye: Nonverbal Thought in Technology,' *Science* 197/4306 (26 August 1977): 834.
19 Robert S. Root-Bernstein, 'Visual Thinking: The Art of Imagining Reality,' *The Visual Arts and Sciences: Transactions of the American Philosophical Society* 75, pt. 6 (1985): 53.
20 Michael Lynch, 'Science in the Age of Mechanical Reproduction: Moral and Epistemic Relations between Diagrams and Photographs,' *Biology and Philosophy* 6/2 (April 1991): 208.
21 Barbara Maria Stafford, *Artful Science: Enlightenment Entertainment and the Eclipse of Visual Education* (Cambridge, MA: MIT Press, 1994), xxii.
22 Ibid., xxv.
23 Edward R. Tufte, *Envisioning Information* (Cheshire, CT: Graphics, 1990), 13.
24 Ulrich Lang and Michel Grave in their article on scientific visualization discuss the visual representation of data describing complex flow systems or financial forecasting. Current visualization programs take advantage of 'the broad bandwidth of the human sensory system' and use graphical displays to map complex and rapidly changing inputs such as world exchange rates. These new methods of displaying data acknowledge the primacy of vision: 'Since vision dominates our sensory input, strong efforts have been made to bring the power of mathematical abstraction and modeling to our eyes through the medium of computer graphics: In Ulrich Lang and Michel Grave, 'Data Structures in Scientific Visualization,' in *Focus on Scientific Visualization*, ed. H. Hagen, H. Müller, and G.M. Nielson (Berlin: Springer-Verlag, 1993), 85.
25 'If we define prints from the functional point of view so indicated, rather than by any restriction of process or aesthetic value, it becomes obvious that without prints we should have very few of our modern sciences, technologies, archaeologies, or ethnologies – for all are dependent, first and last, upon information conveyed by exactly repeatable visual or pictorial statements': William Ivins, Jr., *Prints and Visual Communication* (Cambridge, MA: MIT Press, 1985 [1953]), 3.
26 Martin Rudwick, *The Great Devonian Controversy: The Shaping of Scientific Knowledge among Gentlemen Specialists* (Chicago: University of Chicago Press, 1985), 11–12.
27 A prime example of this practice is the order of works in A.M. Lysaght's catalogue *The Book of Birds* (London: Phaidon, 1975), which mixes periods, styles, and cultures with no discernible order.
28 David Knight, *Zoological Illustration: An Essay Towards a History of Printed Zoological Pictures* (Folkestone: Dawson Archon Books, 1977), 39.

29 Michel Foucault, *The Archaeology of Knowledge*, trans. by A.M. Sheridan Smith (London: Tavistock, 1972), 174.
30 Elizabeth L. Eisenstein, *The Printing Press as an Agent of Change: Communication and Cultural Transformations in Early Modern Europe* (Cambridge: Cambridge University Press, 1979), 113.
31 J.A. Lohne, 'The Increasing Corruption of Newton's Diagrams,' *History of Science* 6 (1967): 72.
32 John R. Levene, 'Sources of Confusion in Descartes's Illustrations, with Reference to the History of Contact Lenses,' *History of Science* 6 (1967): 91–2.
33 Root-Bernstein, 'Visual Thinking,' 62.
34 Gombrich, 'Mirror and Map,' 130.
35 Kemp, 'Taking It on Trust,' 128.
36 Ibid., 129.
37 *Some Memoirs of the Life and Works of George Edwards, Fellow of the Royal and Antiquarian Societies* (London: Printed for J. Robson, Bookseller, 1776), 14.
38 Jules Prown writes, in 'The Truth of Material Culture,' that, 'because underlying cultural asssumptions and beliefs are taken for granted or repressed, they are not visible in what a society says, or does, or makes – its self-conscious expressions. They are, however, detectable in the way things are said, or done, or made – that is, in their style. The analysis of style, I believe, is one key to cultural understanding': From Lubar and Kingery, eds., *History from Things*, 4.
39 Peter J. Taylor and Ann S. Blum. 'Pictorial Representation in Biology,' *Biology and Philosophy* 6/2 (April 1991): 130.
40 James R. Griesemer, 'Must Scientific Diagrams Be Eliminable? The Case of Path Analysis,' *Biology and Philosophy* 6/2 (April 1991): 156.
41 Ibid., 161.
42 Michael Lynch, 'Science in the Age of Mechanical Reproduction: Moral and Epistemic Relations between Diagrams and Photographs,' *Biology and Philosophy* 6/2 (April 1991): 205–26.
43 Gombrich, 'Mirror and Map,' 142.

BIBLIOGRAPHY

Primary Sources

UNPUBLISHED

Images

The prints, drawings, maps, and book illustrations included in this work have been assembled from a number of different collections, the most important of which are noted here:

Blacker-Wood Library of Biology, McGill University, Montreal
Canadian Museum of Nature, Ottawa
Hudson's Bay Company Archives, Provincial Archives of Manitoba, Winnipeg
National Archives of Canada, Ottawa
National Library of Canada, Ottawa
The Natural History Museum, London
New York Botanical Garden, New York
New York Historical Society, New York
Nova Scotia Museum, Halifax
Plant Research Library, Centre for Land and Biological Resource Research, Agriculture Canada, Ottawa

Manuscripts and Sketchbooks

Back, George. Sketchbooks I and II, 1819–22. National Archives of Canada, Ottawa.

Fothergill, Charles. 'Canadian Researches Chiefly in Natural History by C.F. commenced in the Autumn of 1816 ... Materials for A Natural History of the World.' Manuscript journal, 1816–21. Thomas Fisher Rare Book Room, University of Toronto, Toronto.

PUBLISHED

Books

Albin, Eleazar. *A Natural History of Spiders and Other Curious Insects.* London: various publishers, 1736.

Back, George. *Narrative of the Arctic Land Expedition ... in the years 1833, 1834, and 1835* ... London: John Murray, 1836.

Bacon, Francis. *The Essays.* Ed. John Pitcher. Harmondsworth: Penguin, 1987.

Bartram, William. *Observations on the* Inhabitants, Climate, Soil, Rivers, Productions, Animals, *and other matters worthy of Notice ... To which is annex'd, a curious Account of the Cataracts at* Niagara *By Mr. Peter Kalm, A* Swedish *Gentleman who travelled there.* London: J. Whiston and B. White, 1751.

Bauhini, Caspari. *Theatri botanici sive Historiae plantarum* ... Basel: Johannem König, 1658.

Bewick, Thomas. *A History of British Birds: Selections.* London: Paddington, 1976.

Cartier, Jacques. *Jacques Cartier: Relations. Édition critique.* Ed. Michel Bideaux. Montreal: Presses de l'Université de Montréal, 1986.

– *The Voyages of Jacques Cartier.* Ed. Ramsay Cook. Toronto: University of Toronto Press, 1993.

Catesby, Mark. *The Natural History of Carolina, Florida and the Bahama Islands Containing two hundred and twenty figures of Birds, Beasts, Fishes, Serpents, Insects, and Plants.* 1771 ed. facs. Savannah, GA: Beehive, 1974.

Champlain, Samuel de. *Les Voyages de la Nouvelle France occidentale, dicte Canada: Faits par le Sr. de Champlain* ... Paris: Chez Pierre Le-Mvr, 1632.

– *Les Voyages de Samuel Champlain, Saintongeais, Père du Canada.* Ed. Hubert Deschamps. Paris: Presses Universitaires de France, 1951.

– *Les Voyages du Sieur de Champlain Xaintongeois, Capitaine ordinaire pour le Roy ... Journal tres-fidele des observations* ... Paris: Chez Jean Berjon, 1613; Ann Arbor: University Microfilms, 1966.

– *The Works of Samuel de Champlain.* Vol. 2: *1608–1613.* Ed. H.P. Biggar. Toronto: The Champlain Society, 1925.

Cockayne, Sir Thomas. *A Short Treatise of Hunting, 1591,* with an introduction by W.R. Halliday. [Oxford]: Humphrey Milford, Oxford University Press, 1932.

Cornut, Jacques-Philippe. *Canadensium Plantarum aliarúmque nondum editarum Historia* ... Paris, Simonem le Moyne, 1635; Sources of Science, No. 37. New York: Johnson Reprint Corp., 1966.

Cuvier, G.L.C.F.D. *Rapport historique sur les progrès des sciences naturelles depuis 1789, et sur leur état actuel* ... Paris: De l'Imprimerie impériale, 1810. *Spotlights on the History of Science* II. Amsterdam: B.M. Israel, 1970.

Dantiscano, M. Gotardo Arthusio. *Indiae Orientalis, Pars X,* with engravings by Johanne-Theodoro de Bry. Frankfurt, 1613.

de Bry, Johann Theodor, ed. *Dreyzehender Theil Americae, das ist, Fortsetzung der*

Historien von der Newen Welt, oder Nidergängischen Indien, waran aes auff diese Zeit noch anhero ermangelt: darinnen erstlich ein sattsame und gründtliche Beschreibung dess Newen Engellandts ... Franckfurt: Caspar Rötel, 1628.
de Bry, Johann Theodor, and Matthaeus Merian, eds. *Zehender Theil Americae darinnen zubefinden: erstlich, zwo Schiffarten Herrn Americi Vesputii vnter König Ferdinando in Castilien vollbracht: zum andern, ein gründlicher Bericht von dem jetzigen Zustand der Landschafft Virginien* ... Oppenheim: Hieronymo Gallern, 1618.
de Bry, Theodor, ed. *Der ander Theyl, der newlich erfundenen Landtschafft Americae, von dreyen Schiffahrten, so die Frantzosen in Floridam ... gethan* ... Franckfurt am Mayn: Johann Feyerabendendt, 1591.
– *Wunderbarliche, doch warhafftige Erklärung / von der Gelegenheit vnd Sitten der Wilden in Virginia* ... Franckfurt am Mayn: J. Wechel, 1590.
de Charlevoix, P.-F.-X. *Histoire et description générale de la Nouvelle France*. Paris, 1744.
– *Journal of a Voyage to North-America Undertaken by Order of the French King containing The Geographical Description and Natural History of that Country, particularly Canada ... in two volumes*. London: R. and J. Dodsley, 1761; Ann Arbor: University Microfilms, 1966.
de Laet, Johannis. *Novus orbis, seu Descriptiones Indiae Occidentales libri XVIII*. Leyden, 1633.
de Lahontan, Baron Louis-Armand de Lom d'Arce. *New Voyages to North-America. Written in* French *by the Baron Lahontan, Done into* English ... 2 vols. London: H. Benwicke and others, 1703; repr. New York: Reuben Gold Thwaites, Burt Franklin, 1905 [1970].
– *Voyages du Baron La Hontan dans l'Amérique septentrionale, tome premier*. Amsterdam: Chez François l'Honoré, 1705; Montreal: Éditions Elysée, 1974.
– *Mémoires de l'Amérique septentrionale ou la suite des voyages de Mr. Le Baron de La Hontan, tome second*. Amsterdam: Chez François l'Honoré & compagnie, 1705; Montreal: Éditions Elysée, 1974.
Denys, Nicolas. *The Description and Natural History of the Coast of North America (Acadia)*. Trans. and ed. W.F. Ganong. Toronto: The Champlain Society, 1908.
de Pass, Crispin. *Hortus Floridus*, 2 vols. Utrecht: Salomon de Roy, 1615; London: Crosset, 1929.
Du Creux, François. *Historiae canadensis seu Novæ-Franciæ*. Paris: Sebastian Cramoisy & Sebast. Mabre-Cramoisy, 1664.
– *The History of Canada or New France*. Ed. James B. Conacher; trans. Percy J. Robinson. Toronto: The Champlain Society, 1951.
Edwards, George. *A Natural History of Uncommon Birds and Some Other Rare and Undescribed Animals, Quadrupeds, Reptiles, Fishes, Insects, etc. In Seven Parts* ... 4 vols. (includes *Gleanings of Natural History*). London: J. Robson, 1805–6.
Ellis, Henry. *A Voyage to Hudson's Bay by the Dobbs Galley and California* ... London: H. Whitridge, 1748.

Evelyn, John. *The Diary of John Evelyn*, 2 vols. Ed. William Bray. London: J.M. Dent, [1907].

Gerard, John. *The Herbal or General History of Plants*. London, 1633; repr. New York: Dover, 1975.

Gilpin, William. *Three Essays: on Picturesque Beauty; on Picturesque Travel; and on Sketching Landscape, to which is added a poem on Landscape Painting*, 2d ed., London: R. Blamire, 1794.

Gmelin, D. Joanne Georgio. *Flora sibirica sive Historia plantarum sibiriae*. St Petersburg (Petropoli): Academia scientiarum, 1747.

Graham, Andrew. *Observations on Hudson's Bay, 1767–1791*. Ed. Glyndur Williams. London: Hudson's Bay Record Society, 1969.

Grew, Nehemiah. *The Anatomy of Plants, 1682*; Sources of Science, no. 11. New York: Johnson Reprint Corp., 1965.

Gronovius, Joh. Fred. *Flora virginica*. Lugduni Batavorum: Cornelius Haak, 1739; repr. Cambridge, MA: Arnold Arboretum, 1946.

Hearne, Samuel. *A Journey from Prince of Wales's Fort in Hudson's Bay to the Northern Ocean ... in the years 1769, 1770, 1771 & 1772*. Dublin: Printed for P. Byrne and J. Rice, 1796; repr. Toronto: Macmillan, 1958.

Hennepin, Louis. *A New Discovery of a* Vast Country in *America ... With a Continuation: Giving an Account of Attempts of the Sieur* De la Salle *upon the Mines of* St. Barbe, *&c. ... By* L. Hennepin, *now Resident in* Holland. *To which is added, Several* New Discoveries *in* North-America, *not publish'd in the* French *Edition*. London: M. Bentley, J. Tonson, H. Bonwick, T. Goodwin, and S. Manship, 1698.

– *Voyage curieux du RP Louis Hennepin, Missionaire Recollect, & Notaire Apostolique, qui contient Une Nouvelle Decouverte d'un Tres Grand Pays, Situé dans l'Amerique* ... Leide (Leiden): Chez Pierre Vander Aa, 1704.

Houston, C.S., ed. *Arctic Artist: The Journals and Paintings of George Back, Midshipman with Franklin, 1819–1822*. Houston. Montreal: McGill-Queen's University Press, 1994.

– *Arctic Ordeal: The Journal of John Richardson, Surgeon-Naturalist with Franklin, 1820–1822*. Kingston: McGill-Queen's University Press, 1984.

– *To the Arctic by Canoe, 1819–1821: The Journal and Paintings of Robert Hood, Midshipman with Franklin*. Montreal: McGill-Queen's University Press, 1974.

Hunter, Clark, ed. *The Life and Letters of Alexander Wilson*. Philadelphia: American Philosophical Society, 1983.

Isham, James. *James Isham's Observations on Hudson's Bay and Notes and Observations on a book entitled* A Voyage to Hudson Bay in the Dobbs Galley, 1749. London: Hudson's Bay Record Society, 1949.

Kalm, Pehr. *Pehr Kalm's Travels in North America: The English Version of 1770*, 2 vols. Ed. Adolph B. Benson. New York: Dover, 1966.

Kenyon, W.A., ed. *The Journal of Jens Munk, 1618–1620*. Toronto: Royal Ontario Museum, 1980.

Lescarbot, Marc. *The History of New France*, trans. W.L. Grant, with an introduction by H.P. Biggar, 3 vols. Toronto: The Champlain Society, 1914.
Linnaeus, Carolus. *Genera Plantarum, fifth edition, 1754.* With an introduction by William T. Stearn. Historiae naturalis classica 3. Facs. repr. Weinheim, 1960.
– *Hortus Cliffortianus.* Amstelaedami, 1737 (1738).
– *Species Plantarum: A Facsimile of the first edition, 1753.* With an introduction by William T. Stearn. 2 vols. London: The Ray Society, 1959.
– *A System of Vegetables ... Translated from the Thirteenth edition (as published by Dr. Murray) of the Systema Vegetabilium ...* Lichfield: John Jackson for Leigh and Sotheby, London, 1783.
Merian, Maria Sibylla. *Histoire générale des Insectes de Surinam et de Toute l'Europe ...* 3d ed. Ed. M. Buch'oz. Paris: L.C. Desnos, 1771.
– *Leningrad Watercolours.* Facsimile edition, eds. Helen and Kurt Wolff. New York: Harcourt, Brace, Jovanovich, 1974.
Monardes, Nicolas. *Joyfull newes out of the newe founde worlde* ..., 2 vols. Englished by John Frampton, Merchant. Introduction by Stephen Gasellee. New York: AMS, 1967.
Mozino, José Mariano. *Noticias de Nutka: An Account of Nootka Sound in 1792.* Trans. Iris Wilson. Seattle: University of Washington Press, 1970.
Paracelsus. *Selected Writings.* Edited and introduced by Jolande Jacobi, 2d ed. Bollingen Series 28, Princeton, NJ: Princeton University Press, 1969.
Pennant, Thomas. *Arctic Zoology*, 2 vols. London: Henry Hughs, 1784.
– *Arctic Zoology*, 2 vols. London: Robert Faulder, 1792.
– *Beschrijving van de Noorder-Poollanden.* Amsterdam: H. Gratman en W. Vermndel, 1798.
– *Introduction to the Arctic Zoology*, 2d ed. London: Robert Faulder, 1792.
Phébus, Gaston. *Livre de Chasse*, édité avec introduction ... par Gunnar Tilander. Karlshamn: E.G. Johanssons Boktryckeri, 1971.
Pliny the Elder. *Natural History: A Selection.* London: Penguin, 1991.
– *Natural History.* Ed. W.H.S. Jones. Cambridge, MA: Harvard University Press; London: William Heinemann, 1956.
Postels, Aleksandr Filippovich, and F. Ruprecht. *Illustrationes Algarum in itinere circa orbem jussu imperatoris Nicolai I.* St Petersburg, 1840.
Purchas, Samuel. *Henry Hudson's Voyages*, from *Purchas His Pilgrimes.* Ann Arbor, MI: University Microfilms, 1966.
Reynolds, Sir Joshua. *Discourses on Art.* Ed. Robert R. Ware. New Haven, CT, and London: Yale University Press, 1975.
Ross, Sir John. *Narrative of a Second Voyage in Search of a North-West Passage ...* Philadelphia: E.L. Carey & A. Hart, 1835.
Sagard, Gabriel. *The Long Journey to the Country of the Hurons.* Ed. G.M. Wrong; trans. H.H. Langton. Toronto: The Champlain Society, 1939.
Schedel, Hartmann. *The Nuremberg Chronicle: A Facsimile of Hartmann Schedel's Buch der Chroniken, Printed by Anton Koberger in 1493.* New York: Landmark, 1979.

Settle, Dionysus. *A true reporte of the laste voyage into the West and Northwest regions, &c., 1577, worthily atchieved by Capteine Frobisher of the sayde voyage the first finder and Generall. With a description of the people there ihabiting, and other circumstances notable ...* London: Henrie Middleton, 1577; repr. New York: Da Capo, 1969.
Thevet, André. *André Thevet's North America: A Sixteenth-Century View.* Ed. and trans. Robert Schlesinger and A.P. Stabler. Kingston: McGill-Queen's University Press, 1986.
– *La Cosmographie universelle d'André Thevet cosmographe du Roy: Illustree de diverse figures des choses plus remarquables veuës par l'Auteur, & incogneuës de noz Anciens & Modernes.* Paris: Guillaume Chaudiere, 1575.
– *Les Singularitez de la France Antartique: Autrement nommée Amerique: & de plusieurs Terres & Isles decouvertes de nostre temps.* Paris: Maurice de la Porte, 1558.
Topsell, Edward. *The Fowles of Heaven or History of Birdes.* Ed. T.P. Harrison and F. David Hoeniger. Austin, TX: University of Texas Press, 1972.
– *History of Four-Footed Beasts and Serpents and Insects.* Vol. 1: *The History of Four-footed Beasts.* 1658; repr. New York: Da Capo, 1967.
Turberuile, George. *The Booke of Faulconrie or Hauking, for the onely delight and pleasure of all Noblemen and Gentlemen, London 1575.* The English Experience, no. 93. Amsterdam: Theatrum orbis terrarum; New York: Da Capo, 1969.
Turner, William. *De Historia Avium or Turner on Birds: A Short and Succinct History of the Principal Birds Noticed by Pliny and Aristotle.* Ed. A.H. Evans. Cambridge: Cambridge University Press, 1903.
– Libellus de Re Herbaria, *1538 [and]* The Names of Herbes, *1548.* Facsimiles ... London: The Ray Society, 1965.
Vaillant, Sebastien. *Botanicon parisiense, ou Denombrement par ordre alphabetique des Plantes Qui se trouvent aux environs de Paris.* Leiden and Amsterdam: Jean and Herman Verbeck; Balthazar Lakemen, 1727.
White, T.H. ed. and trans. *The Book of Beasts, being a translation of a Latin Bestiary of the 12th century.* London: Jonathan Cape, 1954.
Wood, William. 'Excerpts from *New England's Prospect.*' In *This Incomperable Land: A Book of American Nature Writing,* ed. Thomas J. Lyon. New York: Penguin, 1989.

Pamphlets

Some Memoirs of the Life and Works of George Edwards, Fellow of the Royal and Antiquarian Societies. London: Printed for J. Robson, Bookseller, 1776.

Articles

Davies, Thomas. 'An Account of the Jumping Mouse of Canada. Dipus Canadensis. Read June 6, 1797.' *Transactions of the Linnaean Society* 4 (1798): 155–7.

– 'Description of Menura Superba, A Bird of New South Wales,' *Transactions of the Linnaean Society* 6 (1802): 208–9.
Rackett, Rev. Thomas. 'Description of Some Shells found in Canada.' *Transactions of the Linnaean Society* 13 (1822): 42–4.
Shaw, George. 'Descriptions of the Mus Bursarius and Tubularia Magnifica; from Drawings communicated by Major-General Thomas Davies.' *Transactions of the Linnaean Society* 5 (1800): 227–9.
Smith, James Edward. 'Introductory discourse on the rise and progress of Natural History.' *Transactions of the Linnaean Society*. 1 (1791): 1–55.
Stackhouse, John. 'Observations on preserving Specimens of Plants.' *Transactions of the Linnaean Society* 5 (800): 20–5.

Secondary Sources

PUBLISHED

Reference and Bibliographic Aids

Bertin, Léon, et al. *Larousse Encyclopedia of Animal Life*. London: Hamlyn, 1967 (1971).
Bloomfield, Valerie. *Resources for Canadian Studies in Britain with Some Reference to Europe*, 2d ed. London: British Association for Canadian Studies, 1983.

Books and Articles

Ackerman, James S. 'Early Renaissance "Naturalism" and Scientific Illustration.' In *The Natural Sciences and the Arts: Aspects of Interaction from the Renaissance to the 20th Century. An International Symposium*, ed. Allan Ellenius, 1–17. Stockholm: Almqvist & Wiksell International, 1985.
Adamson, Jeremy Elwell. *Niagara: Two Centuries of Changing Attitudes, 1697–1901*. Washington, DC: Corcoran Gallery of Art, 1985.
Albury, W.R., and D.R. Oldroyd. 'From Renaissance Mineral Studies to Historical Geology, in the Light of Foucault's *The Order of Things*,' *British Journal for the History of Science*, X, 3, no. 36, (November 1977): 187–215.
Alden, John. 'A Note on George Edward's *Natural History of Uncommon Birds*.' *Journal of the Society for the Bibliography of Natural History* 5/2, (1969): 135–6.
Allen, David E. *The Naturalist in Britain: A Social History*. London: Allen Lane, 1976.
Allodi, Mary. *Canadian Watercolours and Drawings in the Royal Ontario Museum*, 2 vols. Toronto: Royal Ontario Museum, 1974.
Alpers, Svetlana. *The Art of Describing: Dutch Art in the Seventeenth Century*. Chicago: University of Chicago Press, 1983.
Altick, Richard D. *The Shows of London: A Panoramic History of Exhibitions, 1600–1862*. Cambridge, MA: The Belknap Press of Harvard University, 1978.

Anker, Jean. *Bird Books and Bird Art.* Copenhagen, 1938; repr. New York: Arno, 1974.

Arber, Agnes. *Herbals, Their Origin and Evolution: A Chapter in the History of Botany, 1470–1670*, 2d ed. Darien, CT: Hafner, 1970.

Archer, Mildred. *Natural History Drawings in the India Office Library.* London: Her Majesty's Stationery Office, 1962.

Ashworth, William B. 'Emblematic Natural History of the Renaissance.' In *Cultures of Natural History*, ed. N. Jardine, J.A. Secord, and E.C. Spary, 17–37. Cambridge: Cambridge University Press, 1996.

– 'The Persistent Beast: Recurring Images in Early Zoological Illustration.' In *The Natural Sciences and the Arts: Aspects of Interaction from the Renaissance to the 20th Century. An International Symposium*, ed. Allen Ellenius, 46–66. Stockholm: Almqvist & Wiksell International, 1985.

Baigrie, Brian S., ed. *Picturing Knowledge: Historical and Philosophical Problems Concerning the Use of Art in Science.* Toronto: University of Toronto Press, 1996.

Barber, Lynn. *The Heyday of Natural History, 1820–1870.* London: Jonathan Cape, 1980.

Barker, Nicholas. *Hortus Eystettensis: The Bishop's Garden and Besler's Magnificent Book.* London: The British Library, 1994.

Barsness, Larry. *The Bison in Art: A Graphic Chronicle of the American Bison.* Fort Worth, TX: Northland Press and the Amon Carter Museum, 1977.

Beckett, R.B. 'Photogenic Drawings.' *Journal of Warburg and Courtauld Institutes* 27 (1964): 342–3.

Beddall, Barbara. 'Scientific Books and Instruments for an Eighteenth-Century Voyage around the World: Antonio Pineda and the Malaspina Expedition.' *Journal of the Society for the Bibliography of Natural History* 9/2 (1979): 95–107.

Béland, Mario, ed. *Painting in Quebec, 1820–1850: Essays.* Quebec: Musée du Québec, 1992.

Bell, Michael. *Painters in a New Land: From Annapolis Royal to the Klondike.* Toronto: McClelland & Stewart, 1973.

Belofsky, Harold. 'Engineering Drawing – A Universal Language in Two Dialects.' *Technology and Culture* 32 (1991): 23–46.

Benson, Keith R. 'Biology's "Phoenix": Historical Perspectives on the Importance of the Organism.' *American Zoologist* 29 (1989): 1067–74.

Berger, Carl. *Science, God and Nature in Victorian Canada.* Toronto: University of Toronto Press, 1983.

Bettex, Albert. *The Discovery of Nature.* New York: Simon & Schuster, 1965.

Bland, David. *A History of Book Illustration: The Illuminated Manuscript and the Printed Book*, 2d ed. London: Faber & Faber, 1969.

Bliss, Donald Percy. *A History of Wood Engraving.* London: Spring Books, 1928 (1964).

Blum, Ann Shelby. *Picturing Nature: American Nineteenth-Century Zoological Illustration.* Princeton, NJ: Princeton University Press, 1993.

Blunt, Wilfrid. *The Art of Botanical Illustration*, 4th ed. London: Collins, 1971.

– *The Compleat Naturalist: A Life of Linnaeus.* London: Collins, 1971.

Boas, Marie. *The Scientific Renaissance, 1450–1630.* New York: Harper Brothers, 1962.

Brandon, William. *New Worlds for Old: Reports from the New World and Their Effect on the Development of Social Thought in Europe, 1500–1800.* Athens: Ohio University Press, 1986.

Bridson, Gavin D. 'From Xylography to Holography: Five Centuries of Natural History illustration.' *Archives of Natural History* 16/2 (1989): 121–41.

Buchanan, Handasyde. *Nature into Art: A Treasury of Great Natural History Books.* London: Weidenfeld, 1979.

Buchanan, Robert. *Life and Adventures of Audubon, the Naturalist.* London: J.M. Dent & Sons, [*ca* 1870].

Bucher, Bernadette. *Icon and Conquest: A Structural Analysis of the Illustrations of de Bry's* Great Voyages. Trans. Basia Miller Gulati. Chicago: University of Chicago Press, 1981.

Calmann, Gerta. *Ehret, Flower Painter Extraordinary: An Illustrated Life.* Oxford: Phaidon, 1977.

Carter, John, and Percy H. Muir. *Printing and the Mind of Man*, 2d ed., revised and enlarged. Munich: Karl Pressler, 1983.

Chiapelli, Fredi. *First Images of America: The Impact of the New World in the Old.* Berkeley: University of California Press, 1976.

Clark, Kenneth. *Landscape into Art.* London: John Murray, 1949; new ed., 1976.

Clarke, Larry R. 'The Quaker Background of William Bartram's View of Nature.' *Journal of the History of Ideas* 46/3 (1985): 435–48.

Clarke, Michael. *The Tempting Prospect: A Social History of English Watercolours.* London: British Museum Publications, 1981.

Clarke, T.H. *The Rhinoceros from Dürer to Stubbs, 1515–1799.* London: Sotheby's Publications, 1986.

Coats, Alice M. *The Treasury of Flowers.* London: Phaidon, 1975

Cohen, I. Bernard. *Album of Science: From Leonardo to Lavoisier, 1450–1800.* New York: Charles Scribner's Sons, 1980.

Comisión Quinto Centenario. *La Real Expedición Botánica á Nueva España, 1787–1803.* Madrid: Real Jardín Botánico CSIC, 1987.

Conway, William M., ed. *The Writings of Albrecht Dürer.* London: Peter Owen, 1958.

Copenhaver, Brian P. 'A Tale of Two Fishes: Magical Objects in Natural History from Antiquity to the Scientific Revolution.' *Journal of the History of Ideas* 52/3 (1991): 373–98.

Corbett, Margery. 'The Engraved Title-Page to John Gerarde's *Herball* or *Generall Historie of Plantes*, 1597.' *Journal of the Society for the Bibliography of Natural History* 8/3 (1977): 223–30.

Crosby, Alfred W. *Ecological Imperialism: The Biological Expansion of Europe, 900–1900*. Cambridge: Cambridge University Press, 1986.

Cumming, W.P., R.A. Skelton, and D.B. Quinn. *The Discovery of North America, 1630–1776*. New York: American Heritage Press, 1971–2.

Cumming, W.P., R.S. Skelton, D.B. Quinn, and G. Wiliams. *The Exploration of North America, 1630–1776*. Toronto: McClelland & Stewart, 1974.

Dacos, Nicole. 'Présents américains à la Renaissance: L'assimilation de l'exotisme.' *Gazette des beaux-arts* 78 (1969): 57–64.

Dahl, Edward. 'The Original Beaver Map – De Fer's 1698 Wall Map of America.' *The Map Collector* 29 (December 1984): 22–6.

Dance, S. Peter. *The Art of Natural History: Animal Illustrators and Their Work*. Woodstock, NY: Overlook, 1976.

Danforth, Susan, and William H. McNeill. *Encountering the New World, 1493 to 1800*. Providence, RI: John Carter Brown Library, 1991.

de Bray, Lys. *The Art of Botanical Illustration: The Classic Illustrators and Their Achievements from 1550 to 1900*. Bromley, Kent: Christopher Helm, 1989.

Debus, Allen G. *Man and Nature in the Renaissance*. Cambridge: Cambridge University Press, 1978.

Deetz, James. *In Small Things Forgotten: The Archaeology of Early American Life*. New York: Doubleday Anchor, 1977.

Dennis, Michael A. 'Graphic Understanding: Instruments and Interpretation in Robert Hooke's *Micrographia*.' *Science in Context* 3/2 (1989): 309–64.

Devèze, Michel. *La vie de la forêt française au XVIe siècle*, 2 vols. Paris: École Pratique des Hautes Études, VIe section; SEVPEN, 1961.

de Virville, Ad. Davy. *Histoire de la botanique en France*. Paris: Société d'Édition d'Enseignement Supérieur, 1954.

Dickenson, Victoria. *First Impressions: European Views of the Natural History of Canada from the 16th to the 19th Century*. Kingston: Agnes Etherington Arts Centre, Queen's University, 1992.

Dickenson, Victoria, with Ed Tompkins. *Treasures: 900 years of the European Presence in Newfoundland and Labrador*. St John's: Newfoundland Museum, 1983.

Dionne, N.-E. *Hennepin: Ses voyages et ses oeuvres*, 2d ed. Montreal: Les Éditions Canadiana, n.d.

Doggett, Rachel, ed. *New World of Wonders: European Images of the Americas, 1492–1700*. Washington, DC: Folger Shakespeare Library, 1992.

Doran, Madeleine. 'On Elizabethan "Credulity," with Some Questions Concerning the Use of the Marvelous in Literature.' *Journal of the History of Ideas* 1/2 (1940): 151–76.

Edwards, Phyllis I. 'The Taxonomic Importance of the Original Drawings of Published Illustrations, with Special Reference to Those of the 17th and 18th Century Related to the Flora of the Cape of Good Hope.' *Journal of the Society for the Bibliography of Natural History* 8/4 (1978): 333–41.

Eisenstein, Elizabeth L. *The Printing Press as an Agent of Change: Communication and Cultural Transformations in Early Modern Europe.* Cambridge: Cambridge University Press, 1979.

Ellenius, Allan, ed. *The Natural Sciences and the Arts: Aspects of Interaction from the Renaissance to the 20th Century. An International Symposium.* Stockholm: Almqvist & Wiksell International, 1985.

Elliott, J.H. *The Old World and the New, 1492–1650.* Cambridge: Cambridge University Press, 1970; Canto ed., 1992.

Engstrand, Iris. *Spanish Scientists in the New World: The Eighteenth-Century Expeditions.* Seattle: University of Washington Press, 1981.

Evans, Brian L. 'Ginseng: Root of Chinese–Canadian Relations.' *Canadian Historical Review* 66/4 (1985): 1–26.

Ewan, Joseph. 'Plant Collectors in America. Backgrounds for Linnaeus.' In *Essays in Biohistory and other contributions presented by friends and colleagues to Frans Verdoorn on the occasion of his 60th birthday*, ed. P. Smit and R.J. Ch. V. ter Laage, 19–54. Utrecht: International Association for Plant Taxonomy, 1970.

– *William Bartram: Botanical and Zoological Drawings, 1756–1788.* Philadelphia: American Philosophical Society, 1968.

Farber, Paul L. 'The Development of Taxidermy and the History of Ornithology.' *Isis* 68/1244 (1977): 550–66.

Ferguson, Eugene S. 'Elegant Inventions: The Artistic Component of Technology.' *Technology and Culture* 19 (1978): 450–60.

– 'The Mind's Eye: Nonverbal Thought in Technology.' *Science* 197/4306 (26 August 1977): 827–36.

Flinden, Paula. *Possessing Nature: Museums, Collecting, and Scientific Culture in Early Modern Italy.* Berkeley: University of California Press, 1994.

Ford, Brian J. *Images of Science: A History of Scientific Illustration.* London: The British Library, 1992.

Forsyth, Adrian. *Mammals of the Canadian Wild.* Camden East, ON: Camden House, 1985.

Foucault, Michel. *The Archaeology of Knowledge.* Trans. A.M. Sheridan Smith. London: Tavistock, 1972.

– *The Order of Things: An Archaeology of the Human Sciences.* New York: Vintage, 1973.

Fournier, P. *Les voyageurs naturalistes du clergé français avant la Révolution.* Paris: Paul Lechevalier & Fils, 1932.

Franklin, Wayne. *Discoverers, Explorers, Settlers: The Diligent Writers of Early America.* Chicago: University of Chicago Press, 1979.

Freedberg, David, and Jan de Vries, eds. *Art in History, History in Art: Studies in Seventeenth-Century Dutch Culture.* Santa Monica, CA: The Getty Center for the History of Art and the Humanities, 1991.

Freedberg, David. 'Science, Commerce and Art: Neglected Topics at the Junction of History and Art History.' In *Art in History, History in Art: Studies in Seventeenth-*

Century Dutch Culture, ed. David Freeberg and Jan de Vries. Santa Monica, CA: The Getty Center for the History of Art and the Humanities, 1991.

Gagnon, François-Marc. *Jacques Cartier et la découverte du Nouveau Monde.* Quebec: Musée du Québec, 1984.

Gagnon, François-Marc, and Denise Petel. *Hommes effarables et bestes sauvaiges: Images du Nouveau-Monde d'après les voyages de Jacques Cartier.* Montreal: Les Éditions du Boréal Express, 1986.

Ganong, W.F. *Crucial Maps in the Early Cartography and Place-Nomenclature of the Atlantic Coast of Canada.* Toronto: University of Toronto Press, 1964.

– 'The Identity of Animals and Plants Mentioned by the Early Voyagers to Eastern Canada and Newfoundland.' *Transactions of the Royal Society of Canada*, section II, 3rd series, vol. III (1910): 197–242.

George, Wilma. *Animals and Maps.* Berkeley: University of California Press, 1969.

– 'The Bestiary: A Handbook of the Local Fauna,' *Archives of Natural History* 10/2 (1981): 187–203.

– 'The Living World of the Bestiary.' *Archives of Natural History* 12/1 (1985): 161–4.

– 'Sources and Background to Discoveries of New Animals in the Sixteenth and Seventeenth Centuries.' *History of Science* 18 (1980): 79–104.

Goetzmann, William H., and Antonello, Gerbi. *The Dispute of the New World: The History of a Polemic, 1750–1900.* Revised and enlarged ed., translated by Jeremy Moyle. Pittsburgh: University of Pittsburgh Press, 1973.

Glaser, Lynn. *America on Paper: The First Hundred Years.* Philadelphia: Associated Antiquarians, 1989.

Glassie, Henry. *Folk Housing in Middle Virginia.* Knoxville: University of Tennessee Press, 1975.

Goetzmann, William H. *New Lands, New Men: America and the Second Great Age of Discovery.* New York: Viking, 1986.

– 'Seeing and Believing: The Explorer and the Visualization of Space.' In *Essays on the History of North American Discovery and Exploration*, ed. S. Palmer and D. Reinhartz, 133–40. Dallas: University of Texas A & M Press, 1988.

Gombrich, E.H. *Art and Illusion: A Study in the Psychology of Pictorial Representation*, 2d ed. Princeton, NJ: Princeton University Press, 1972

– *The Heritage of Apelles: Studies in the art of the Renaissance.* London: Phaidon, 1976.

– 'Mirror and Map: Theories of Pictorial Representation.' *Philosophical Transactions of the Royal Society* 270/903 (1974): 119–49.

Goody, Jack. *The Culture of Flowers.* Cambridge: Cambridge University Press, 1993.

Goss, John. *The Mapping of North America.* Secaucus, NJ: Wellfleet, 1990.

Greenblatt, Stephen. *Marvelous Possessions: The Wonder of the New World.* Chicago: University of Chicago Press, 1991.

Greene, Edward Lee. *Landmarks of Botanical History, Part II.* Ed. Frank N. Edgerton. Stanford, CA: Stanford University Press, 1983.

Griesemer, James R. 'Must Scientific Diagrams Be Eliminable? The Case of Path Analysis.' *Biology and Philosophy* 6/2 (April 1991): 155–80.

Hagen, H., H. Müller, G.M. Nielson, eds. *Focus on Scientific Visualization.* Berlin: Springer-Verlag, 1993.

Hall, Bert S. 'The Didactic and the Elegant: Some Thoughts on Scientific and Technological Illustrations in the Middle Ages and Renaissance.' In *Picturing Knowledge: Historical and Philosophical Problems Concerning the Use of Art in Science*, ed. Brian S. Baigrie, 3–39. Toronto: University of Toronto Press, 1996.

Harper, J. Russell. *Thomas Davies in Early Canada.* Ottawa: Oberon, 1972.

Harris, Neil. *Cultural Excursions: Marketing Appetites and Cultural Tastes in Modern America.* Chicago and London: University of Chicago Press, 1993.

Haskell, Francis. *History and Its Images: Art and the Interpretation of the Past.* New Haven, CT: Yale University Press, 1993.

Hays, H.R. *Birds, Beasts, and Men: A Humanist History of Zoology.* Baltimore, MD: Penguin, 1972.

Henry, John Frazier. *Early Maritime Artists of the Pacific Northwest Coast, 1741–1841.* Vancouver, BC: Douglas, 1984.

Heuvelmans, Bernard. *On the Track of Unknown Animals.* Trans. and abridged by Richard Garnett. Cambridge, MA: MIT Press [1955/1962], 1972.

Holt, Elizabeth G., ed. *A Documentary History of Art.* Vol. 2: *Michelangelo and the Mannerists: The Baroque and the Eighteenth Century.* Princeton, NJ: Princeton University Press, 1982.

Honour, Hugh. *The European Vision of America: A Special Exhibition to Honor the Bicentennial of the United States ...*, Cat. no. 92. Cleveland: Cleveland Museum of Art, 1975.

– *The New Golden Land: European Images of America from the Discoveries to the Present Time.* New York: Pantheon, 1975.

Houghton, Walter E., Jr. 'The English Virtuoso in the Seventeenth Century (II).' *Journal of the History of Ideas* 3/2 (1942): 190–219.

Hubbard, R.H., ed. *Thomas Davies, c. 1737–1812.* Ottawa: National Gallery of Canada,1972.

Hulton, Paul B. *America 1585: The Complete Drawings of John White.* Raleigh: University of North Carolina Press, 1984.

– 'Realism and Tradition in Ethnological and Natural History Imagery of the 16th Century.' In *The Natural Sciences and the Arts: Aspects of Interaction from the Renaissance to the 20th Century. An International Symposium*, ed. Allan Ellenius, 18–31. Stockholm: Almqvist & Wiksell International, 1985.

– *The Work of Jacques Le Moyne de Morgues: A Huguenot Artist in France, Florida and England*, 2 vols. London: British Museum Publications, 1977.

Hulton, Paul B., and D.B. Quinn. *The American Drawings of John White, 1577–1590.* London: British Museum Publications, 1964.

Hutchinson, G. Evelyn. 'Aposematic Insects and the Master of the Brussels Initial.' *American Scientist*, March/April 1974, 161–71.

– 'Attitudes toward Nature in Medieval England: The Alphonso and Bird Psalters.' *Isis* 65 (1974): 5–29.

Inkster, Ian, and Jack Morrell. *Science in British Culture, 1780–1845*. London: Hutchinson, 1983.

Ivins, William M., Jr. *On the Rationalization of Sight, with an Examination of the Renaissance Texts on Perspective*. Metropolitan Museum of Art Papers no. 8. New York: Metropolitan Museum of Art, 1938.

– *Prints and Visual Communication*. Cambridge, MA: MIT Press, 1985 (1953).

Jackson, C.E. *Bird Illustrators: Some Artists in Early Lithography*. London: H.F. & G. Witherby, 1975.

Jenkins, Alan C. *The Naturalists: Pioneers of Natural History*. London: Hamish Hamilton, 1978.

Jones, Howard Mumford. *O Strange New World. American Culture: The Formative Years*. New York: Viking, 1964.

Kastner, Joseph. *A Species of Eternity*. New York: Alfred A. Knopf, 1977.

Kemp, Martin. *The Science of Art. Optical Themes in Western Art from Brunelleschi to Seurat*. New Haven, CT: Yale University Press, 1990.

– 'Taking It on Trust: Form and Meaning in Naturalistic Representation.' *Archives of Natural History* 17/2 (1990): 127–88.

– 'Temples of the Body and Temples of the Cosmos: Vision and Visualization in the Vesalian and Copernican Revolutions.' In *Picturing Knowledge: Historical and Philosophical Problems Concerning the Use of Art in Science*, ed. Brian S. Baigrie, 40–85. Toronto: University of Toronto Press, 1996.

Klingender, Francis. *Animals in Art and Thought to the End of the Middle Ages*. London: Routledge & Kegan Paul, 1971.

Knight, D.M. 'Background and Foreground: Getting Things in Context.' *British Journal for the History of Science* 20 (1987): 3–12.

– *Natural Science Books in English, 1600–1900*. London: B.T. Batsford, 1972

– *Zoological Illustration: An Essay towards a History of Printed Zoological Pictures*. Folkestone: Dawson Archon Books, 1977.

Krohn, Roger. 'Why Are Graphs So Central in Science?' *Biology and Philosophy* 6/2 (April 1991): 181–203.

Kubler, George. *The Shape of Time: Remarks on the History of Things*. New Haven, CT: Yale University Press, 1962.

Lamontagne, Roland. *La Galissonière et le Canada*. Montreal: Presses de l'Université de Montréal, 1962.

Land, Ulrich, and Michel Grave. 'Data Structures in Scientific Visualization.' In *Focus on Scientific Visualization*, ed. H. Hagen, H. Müller, and G.M. Nielson. Berlin: Springer-Verlag, 1993.

Lane, Christopher W. *Impressions of Niagara: The Charles Rand Penney Collection of Prints of Niagara Falls and the Niagara River from the Sixteenth to the Early Twentieth Century*. Philadelphia: The Philadelphia Print Shop, 1993.

Largen, M.F., and V. Rogers-Price. 'John Abbott, an Early Naturalist-Artist in

North America: His Contributions to Ornithology, with Particular Reference to a Collection of Bird Skins in the Merseyside County Museum, Liverpool.' *Archives of Natural History* 12/2 (1985): 231–52.

Leighton, Ann. *American Gardens in the Eighteenth Century: 'For Use or For Delight.'* Amherst: University of Massachusetts Press, 1986.

Leith-Ross, Prudence. *The John Tradescants: Gardeners to the Rose and Lily Queen.* London: Peter Owen, 1984.

Lemon, Donald P. *Theatre of History: Three Hundred Years of Maps of the Maritimes.* Saint John: New Brunswick Museum Publications, 1987.

Leroy, Jean, ed. *Les botanistes français en Amérique du nord avant 1850.* Colloques internationaux du Centre National de la recherche, LXIII. Paris: Centre National de la Recherche Scientifique, 1957.

Levene, John R. 'Sources of Confusion in Descartes's Illustrations, with Reference to the History of Contact Lenses.' *History of Science* 6 (1967): 90–6

Levenson, Jay A., ed. *Circa 1492: Art in the Age of Exploration,* Cat no. 205. Washington, DC: National Gallery of Art; New Haven, CT, and London: Yale University Press, 1991.

Levere, Trevor. *Science and the Canadian Arctic: A Century of Exploration, 1818–1918.* Cambridge: Cambridge University Press, 1993.

Levere, Trevor, and R.H. Jarrell. *A Curious Field-Book: Science and Society in Canadian History.* Toronto: Oxford University Press, 1974.

Lloyd, Clare. *The Travelling Naturalists.* London: Croom Helm, 1985.

Lohne, J.A. 'The increasing corruption of Newton's Diagrams.' *History of Science* 6 (1967): 69–89.

'Long Lost Sessé and Mociño Illustrations Acquired.' *Bulletin of the Hunt Institute for Botanical Documentation* 3/1 (Spring/Summer 1981): 1–2.

Loomis, Chauncey C. 'The Arctic Sublime.' In *Nature and the Victorian Imagination,* ed. U.C. Knoepflmacher and G.B. Tennyson, 95–112 Berkeley: University of California Press, 1977.

Lubar, Stephen, and W.D. Kingery, eds. *History from Things: Essays on Material Culture.* Washington, DC: Smithsonian Institution Press, 1993.

Lynch, Michael. 'Science in the Age of Mechanical Reproduction: Moral and Epistemic Relations between Diagrams and Photographs.' *Biology and Philosophy* 6/2 (April 1991) 205–26.

Lysaght, A.M. *The Book of Birds.* London: Phaidon, 1975.

– *Joseph Banks in Newfoundland & Labrador, 1766: His Diary, Manuscripts and Collections.* Berkeley: University of California Press, 1971.

Mabey, Richard. *The Flowering of Kew: 200 Years of Flower Paintings from the Royal Botanic Gardens.* London: Century, 1988.

– *Gilbert White. A Biography of the Author of* The Natural History of Selborne. London: Century, 1986.

Macgregor, Arthur. 'Animals and the Early Stuarts: Hunting and Hawking at the Court of James I and Charles I.' *Archives of Natural History* 16/3 (1989): 305–18.

Manning, Roger B. *Hunters and Poachers: A Social and Cultural History of Unlawful Hunting in England, 1485–1640.* Oxford: The Clarendon Press, 1993.

Manthorne, Katherine E. *Tropical Renaissance: North American Artists Exploring Latin America, 1839–1879.* Washington, DC: Smithsonian Institution Press, 1989.

Maquet, Jacques. 'Objects as Instruments, Objects as Signs.' In *History from Things: Essays on Material Culture*, ed. S. Lubar and W.D. Kingery, 30–40. Washington, DC: Smithsonian Institution Press, 1993.

Mason, A. Stuart. *George Edwards: The Bedell and His Birds.* London: Royal College of Physicians, 1992.

Mayor, Hyatt. *Prints and People: A Social History of Printed Pictures.* New York: Metropolitan Museum of Art, 1971.

McAtee, W. 'North American Bird Records in the "Philosophical Transactions," 1665–1800.' *Journal of the Society for the Bibliography of Natural History* 3/1 (1953): 56–60.

– 'North American Birds of George Edwards.' *Journal of the Society for the Bibliography of Natural History* 2/5 (1950): 194–205.

– 'North American Birds of Linnaeus.' *Journal of the Society for the Bibliography of Natural History* 3/5 (1957): 291–300.

– 'North American Birds of Mark Catesy and Eleazar Albin.' *Journal of the Society for the Bibliography of Natural History* 3/4 (1957): 177–94.

– 'North American Birds of Thomas Pennant.' *Journal of the Society for the Bibliography of Natural History* 4/2 (1963): 100–24.

– 'North American Birds of the Virginia Chroniclers, 1588–1686.' *Journal of the Society for the Bibliography of Natural History* 3/2 (1955): 92–101.

McGee, Harold. *On Food and Cooking: The Science and Lore of the Kitchen.* New York: Collier's, 1984.

McManus, Douglas R. *European Impressions of the New England Coast, 1497–1620,* Research paper no. 139. Chicago: University of Chicago, Department of Geography, 1972.

Medsger, Oliver P. *Edible Wild Plants.* 1939, repr. New York: Collier's, 1976.

Merrill, Lynn L. *The Romance of Victorian Natural History.* New York: Oxford University Press, 1989.

Montreal Museum of Fine Arts. *The Painter and the New World.* Montreal: Museum of Fine Arts, 1967.

Moore, John A. *Science as a Way of Knowing: The Foundations of Modern Biology.* Cambridge, MA: Harvard University Press, 1993.

Morison, Samuel Eliot. *The European Discovery of America: The Northern Voyages, AD 500–1600.* New York: Oxford University Press, 1971.

Morison, W.K. 'The Procupine [*sic*] Map.' *ACML Bulletin* 62 (March 1987): 18.

Morris, P.A. 'An Historical Review of Bird Taxidermy in England.' *Archives of Natural History* 20/2 (1993): 241–55.

Mules, Helen B. *Flowers in Books and Drawings, ca 940–1840.* New York: J. Pierpont Morgan Library, 1980.

Museum of Fine Arts, Boston. *Albrecht Duerer, Master Printmaker.* Boston: Museum of Fine Arts, 1971.
National Library of Canada. *Images of Flora and Fauna/Images de la Flore et de la Faune.* Ottawa: National Archives of Canada, 1989.
– *Passages. A Treasure Trove of North American Exploration/Passages. Un écrin des explorations de l'Amérique du Nord.* Ottawa: National Library of Canada, 1992.
Norelli, Martina R. *American Wildlife Painting.* New York: Watson-Guptill, 1975.
Nova Scotia Museum. *Five Centuries of Botanical Illustration.* Halifax: Nova Scotia Museum, n.d.
Novak, Barbara. *Nature and Culture: American Landscape and Painting, 1825–1875.* New York: Oxford University Press, 1980.
Nygren, Edward. *Views and Visions: American Landscape before 1830.* Washington, DC: Corcoran Gallery of Art, 1986.
Pacht, Otto. 'Early Italian Nature Studies and the Early Calendar Landscape.' *Journal of the Warburg and Courtauld Institutes* 1/13 (1950): 13–47.
Panofsky, Erwin. 'Galileo as a Critic of the Arts: Aesthetic Attitude and Scientific Thought.' *Isis,* 47/1 (1956): 3–15.
Pastoureau, Mireille. *Voies océanes de l'ancien au nouveau monde.* Paris: Éditions Hérvas, 1990.
Peattie, Donald C. *A Natural History of Trees of Eastern and Central North America.* 1948; repr. Boston: Houghton Mifflin, 1991.
Pietsch, Theodore W. 'Louis Renaud's Fanciful Fishes.' *Natural History* 93/1 (January 1984): 58–67.
Porter, Charlotte. 'The Drawings of William Bartram (1739–1823), American naturalist.' *Archives of Natural History* 16/3 (1989): 289–303.
Prest, John. *The Garden of Eden: The Botanic Garden and the Re-Creation of Paradise.* New Haven, CT, and London: Yale University Press, 1981.
Pringle, James S. 'How "Canadian" Is Cornut's Canadensium Plantarum Historia? A Phytogeographic and Historical Analysis.' *Canadian Horticultural History* 1/4 (1988): 190–209.
Prioul, Didier. 'British Landscape Artists in Quebec: From Documentary Views to a Poetic Vision.' In *Painting in Quebec, 1820–1850,* ed. Mario Béland, 50–9. Quebec: Musée du Québec, 1992.
Prown, Jules. 'The Truth of Material Culture: History or Fiction?' In *History from Things: Essays on Material Culture,* ed. S. Lubar and W.D. Kingery, 1–19. Washington, DC: Smithsonian Institution Press, 1993.
Rabb, Theodore K., and Jonathan Brown. 'The Evidence of Art: Images and Meaning in History.' *Journal of Interdisciplinary Studies* 17/1 (Summer 1986): 1–6.
Rackham, Oliver. *A History of the Countryside.* London: J.M. Dent & Sons, 1986.
Raven, C.E. *English Naturalists from Neckham to Ray: A Study of the Making of the Modern World.* Cambridge: Cambridge University Press, 1947.
– *John Ray, Naturalist. His Life and Works.* 2d ed. Cambridge: Cambridge University Press, 1950; repr. Cambridge Science Classics, 1986.
Rawson, Jessica, ed. *Animals in Art.* London: British Museum Publications, 1977.

Reed, Jacques. 'Knowledge of the Territory.' *Science in Context* 4/1 (Spring 1991): 133–62.
Reeds, Karen. 'Renaissance Humanism and Botany.' *Annals of Science* 33 (1976): 519–42.
Rhodes, Lynette I. *Science within Art.* Cleveland: Cleveland Museum of Art, 1980.
Riddle, John M. 'Pseudo-Dioscorides' *Ex herbis femininis* and Early Medieval Medical Botany.' *Journal of the History of Biology* 14/1 (Spring 1981): 43–81.
Rix, Martyn. *The Art of the Botanist.* Guildford: Lutterworth, 1981.
Robin, Harry. *The Scientific Image: From Cave to Computer.* New York: Harry N. Abrams, 1992.
Roe, F.G. *The North American Buffalo: A Critical Study of the Species in Its Wild State*, 2d ed. Toronto: University of Toronto Press, 1970.
Rolfe, W.D. Ian. 'William Hunter (1718–1783) on Irish "elk" and Stubbs's Moose.' *Archives of Natural History* 11/2 (1983): 263–90.
Rookmaker, L.C. 'Two Collections of Rhinoceros Plates Compiled by James Douglas and James Parsons in the Eighteenth Century.' *Journal of the Society for the Bibliography of Natural History* 9/1 (1978): 17–38.
Root-Bernstein, Robert S. 'Visual Thinking: The Art of Imagining Reality.' *The Visual Arts and Sciences: Transactions of the American Philosophical Society* 75, pt. 6 (1985): 50–67.
Rousseau, Jacques. 'Quelques jalons dans l'histoire de la botanique de la Nouvelle France, de Cartier à la fin de régime français. In *Les botanistes français en Amérique du nord avant 1850*, ed. Jean Leroy, 150–7. Paris: Centre National de la Recherche Scientifique, 1958.
Rudwick, Martin J.S. 'The Emergence of a Visual Language for Geological Science, 1760–1840.' *History of Science* 14 (1976): 149–95.
– *The Great Devonian Controversy: The Shaping of Scientific Knowledge among Gentlemen Specialists.* Chicago: University of Chicago Press, 1985.
– *The Meaning of Fossils: Episodes in the History of Palaeontology.* London: Macdonald, 1972.
Ruestow, Edward G. 'Images and Ideas: Leeuwenhoek's Perception of the Spermatazoa.' *Journal of the History of Biology* 16/1 (Summer 1983): 185–224.
Ryan, Michael T. 'Assimilating New Worlds in the Sixteenth and Seventeenth Centuries.' *Comparative Studies in Society and History* 23 (1981): 519–38.
Santillana, George. 'The Role of Art in the Scientific Renaissance.' In *Critical Problems in the History of Science*, ed. M. Clagett, 33–65. Madison: University of Wisconsin Press, 1962.
Sawyer, F.C. 'Notes on Some Original Drawings of Birds Used by Dr. John Latham.' *Journal of the Society for the Bibliography of Natural History* 2/5 (1949): 173–80.
– 'A Short History of the Libraries and List of MSS. and Original Drawings in the British Museum (Natural History).' *Bulletin of the British Museum (Natural History), Historical Series* 4/2 (1971).
– 'Some Natural History Drawings Made during Captain Cook's First Voyage

round the World.' *Journal of the Society for the Bibliography of Natural History* 2/5 (1949): 190–3.

Schaffer, Simon. 'Herschel in Bedlam: Natural History and Stellar Astronomy.' *British Journal for the History of Science*, XIII, 3, no. 45 (November 1980): 211–39.

Schlereth, T.J. *Cultural History and Material Culture. Everyday Life, Landscapes and Museums*. Ann Arbor, MI: UMI Research Press, 1989.

Schmidt, Karl P. 'The "Methodus" of Linnaeus, 1736.' *Journal of the Society for the Bibliography of Natural History* 2 (1943–52): 369–74.

Sehm, Gunter G. 'The First European Bison Illustration and the First Central European Exhibit of a Living Bison.' *Archives of Natural History* 18/3 (1991): 323–32.

Shirley, John W., and F. David Hoeniger, eds. *Science and Arts in the Renaissance*. Washington, DC: Folger Shakespeare Library, 1984.

Singer, Charles. *From Magic to Science: Essays on the Scientific Twilight*. New York: Dover, 1958.

Sitwell, S., H. Buchanan, and J. Fisher. *Fine Bird Books, 1700–1900*. New York: Atlantic Monthly Press, 1990.

Skelton, R.A. *Explorers' Maps: Chapters in the Cartographic Record of Geographical Discovery*. London: Routledge & Kegan Paul, 1960.

Skipwith, Peyton. *The Great Bird Illustrators and Their Art, 1730–1930*. New York: A&W Publishers, 1979.

Smith, Bernard. *The Art of Captain Cook's Voyages*, 3 vols. in 4°. Melbourne: Oxford University Press, 1985.

– *European Vision and the South Pacific*, 2d ed. New Haven, CT, and London: Yale University Press, 1988.

Sparling, Mary. *Great Expectations: The Euopean Vision in Nova Scotia, 1749–1848* (Halifax, NS): Mount Saint Vincent University Art Gallery, 1980.

Stafleu, Frans A. *Linnaeus and the Linnaeans: The Spreading of Their Ideas in Systematic Botany, 1735–1789*. Utrecht: Oostnoek's Uitgenersmaatschappi N.V., 1971.

Stafford, Barbara Maria. *Artful Science: Enlightenment Entertainment and the Eclipse of Visual Education*. Cambridge, MA: MIT Press, 1994.

– *Voyage into Substance: Art and Science and the Illustrated Travel Account, 1760–1840*. Cambridge, MA: MIT Press, 1984.

Stearn, William T. 'Linnean Classification, Nomenclature, and Method.' In *The Compleat Naturalist: A Life of Linnaeus*, ed. W. Blunt, 242–53. London: Collins, 1971.

Stearns, Raymond P. *Science in the British Colonies of America*. Urbana: University of Illinois Press, 1970.

Sturtevant, William C. 'The Sources of European Imagery of Native Americans.' In *New World of Wonders: European Images of the Americas, 1492–1700*, ed. Rachel Doggett. Washington, DC: Folger Shakespeare Library, 1992.

Taylor, Basil. *Animal Painting in England from Barlow to Landseer.* Harmondsworth: Penguin, 1955.

Taylor, Peter J., and Ann S. Blum. 'Pictorial Representation in Biology.' *Biology and Philosophy* 6/2 (April 1991): 125–34.

Thomas, Keith. *Man and the Natural World: Changing Attitudes in England, 1500–1800.* Harmondsworth: Penguin, 1984.

Thorndike, Lynn. *A History of Magic and Experimental Science.* Vol. 6: *The Sixteenth Century.* New York: Columbia University Press, 1961. Vols. 7 and 8: *The Seventeenth Century.* New York: Columbia University Press, 1958.

Topper, David. 'Natural Science and Visual Art: Reflections on the Interface.' In *Beyond History of Science: Essays in Honor of Robert E. Scholfield*, ed. E. Garber, 296–310. London/Toronto: Associated University Presses, [*ca* 1990].

– 'Towards an Epistemology of Scientific Illustration.' In *Picturing Knowledge: Historical and Philosophical Problems Concerning the Use of Art in Science*, ed. Brian S. Baigrie, 215–49. Toronto: University of Toronto Press, 1996.

Tufte, Edward. *Envisioning Information.* Cheshire, CT: Graphics, 1990.

Turner, James. '*Landscape* and the 'Art Prospective' in England, 1584–1660.' *Journal of Warburg and Courtauld Institutes* 42 (1979): 290–4.

Tyson, G., and S. Wagonheim. *Print and Culture in the Renaissance: Essays on the Advent of Printing in Europe.* Newark: University of Delaware Press, 1986.

Vachon, Auguste. 'Flora and Fauna: Louis Nicolas and the "Codex canadiensis."' *The Archivist* 12/2 (March /April 1985): 1–2.

Valiant, Sharon. 'Questioning the Caterpillar.' *Natural History* 101/12 (December 1992): 46–59.

Vancouver Maritime Museum. *Enlightened Voyages: Malaspina and Galliano on the Northwest Coast, 1791–1792.* Vancouver: Vancouver Maritime Museum, 1990.

Vancouver World Expo. *To the Totem Shore: The Spanish Presence on the Northwest Coast.* Vancouver, 1986.

Vandervell, Anthony, and Charles Coles. *Game and the Engligh Landscape: The Influence of the Chase on Sporting Art and Scenery.* London: Debrett's Peerage, 1980.

Villanueva, Pabellón. *La Botánica en la Expedición Malaspina.* Colección Encuentros. Madrid: Turner, 1989.

Walsh, John. 'Skies and Reality in Dutch Landscaping.' In *Art in History, History in Art: Studies in Seventeenth-Century Dutch Culture*, ed. David Freeberg and Jan de Vries, 95–118. Santa Monica, CA: The Getty Center for the History of Art and the Humanities, 1991.

Wheeler, Alyne, P.S. Davis, and E. Lazenby. 'William Turner's (c 1508–1568) Notes on Fishes in His Letter to Conrad Gesner.' *Archives of Natural History* 13/3 (1986): 291–305.

White, Lynn, Jr. 'Natural Science and Naturalistic Art in the Middle Ages.' *American Historical Review* 52/3 (April 1947): 421–35.

Whitehead, P.J.P. 'The Original Drawings for the *Historia Naturalis Brasiliae* of Piso and Marcgrave (1648).' *Journal of the Society for the Bibliography of Natural History* 7/4 (1976): 409–22.

Willis, R.J. 'The Earliest Known Australian Bird Painting: A Rainbow Lorikeet, *Trichoglossus haemarodus moluccanus* (Gmelin), by Moses Griffiths, painted in 1772.' *Archives of Natural History* 15/3 (1988): 323–9.

Wilson, Catherine. 'Visual Surface and Visual Symbol: The Microscope and the Occult in Early Modern Science.' *Journal of the History of Ideas* 49/1 (1988): 85–108.

Winkler, Mary G., and Albert Van Helden. 'Representing the Heavens: Galileo and Visual Astronomy.' *Isis* 83 (1992): 195–217.

Wittkower, Rudolf. 'Miraculous Birds.' *Journal of Warburg and Courtauld Institutes* 1 (1937–8; repr. 1965): 253–7.

Zeller, Suzanne. *Inventing Canada: Early Victorian Science and the Idea of a Transcontinental Nation.* Toronto: University of Toronto Press, 1989.

– 'The Spirit of Bacon: Science and Self-Perception in the Hudson's Bay Company, 1830–1870.' *Scientia Canadensis* 13/2 (Fall/Winter 1989): 79–101.

Zetterberg, J. Peter. 'Echoes of Nature in Salomon's House.' *Journal of the History of Ideas* 43/2 (1982): 179–94.

CREDITS

Author: Plate 1; Plate 35

Blacker-Wood Library of Biology, McGill University, Montreal: Plate 36; Plate 37; Plate 38; Plate 39

Canadian Museum of Nature, Ottawa: Plate 19 (CMN S92–002); Plate 48

Hudson's Bay Company Archives, Provincial Archives of Manitoba, Winnipeg: Plate 34 (HBCA P-318)

National Archives of Canada, Ottawa: Plate 2 (NMC-40461); Plate 3 (NMC-52408); Plate 14 (NMC-97952); Plate 21; Plate 22 (Documentary Art and Photography Division [hereafter DAPD], 1992-466-6X; neg. C-014586); Plate 23 (DAPD, 1992-416-1X; neg. C-117226); Plate 43 (DAPD, 1983-91-1; neg. C-105601); Plate 46 (DAPD, 1992-498-1X; neg. C-038858); Plate 52 (DAPD, 1970-188-62; neg. C-40985), W.H. Coverdale Collection of Canadiana; Plate 54 (DAPD, C-141488, 1994-254-2); Plate 55 (DAPD, C-40360; 1970-188-1275); Plate 56 (DAPD, C-141518, 1994-254-2); Plate 57 (DAPD, C-141432, 1994-254-1.41R). Plates 54, 56, and 57 acquired with the assistance of Hoechst and Celanese Canada, with a grant from the Ministry of Heritage under the Cultural Property Import and Export Act.

National Library of Canada. Bibliothèque nationale du Canada, Ottawa: Plate 4 (RBC: NL18054); Plate 5 (RBC: NL18790); Plate 6 (RBC: NL18762); Plate 7 (RBC: NL18763); Plate 8; Plate 9 (RBC: NL18764); Plate 10 (RBC:NL 18196); Plate 11 (RBC: C79478); Plate 12 (RBC: NL18770); Plate 13 (RBC: NL18791); Plate 20; Plate 24 (RBC: NL18767); Plate 25 (RBC: NL18765); Plate 26 (RBC: NL18046); Plate 27 (RBC: NL18049); Plate 28 (RBC: NL 18063); Plate 29 (RBC: NL18768); Plate 30 (RBC: NL18059); Plate 31 (RBC: NL18056); Plate 32 (RBC: NL18058); Plate 33 (RBC: NL18045); Plate 44 (RBC: NL18041); Plate 49 (RBC: NL18766); Plate 50 (RBC: NL18048); Plate 53 (RBC: NL18039)

The Natural History Museum, London: Plate 42

New York Botanical Garden, Bronx, New York: Plate 51

New York Historical Society, New York: Plate 47, Travellers Fund, J.S. Cushman, and Foster-Jarvis Fund 1954 (1954-2)

History Collection, Nova Scotia Museum, Halifax: Plate 45 (NSM 80.11; neg N-14,638)

Plant Research Library CLBRR, Agriculture Canada, Ottawa: Plate 15; Plate 16; Plate 17; Plate 18; Plate 40; Plate 41

INDEX